Transport Properties and Concrete Quality

Transport Properties and Concrete Quality

Materials Science of Concrete,
Special Volume

Proceedings of the Transport Properties and Concrete Quality Workshop, Arizona State University, Tempe, AZ (2005)

Editors
Barzin Mobasher
Jan Skalny

A JOHN WILEY & SONS, INC., PUBLICATION

Published by John Wiley & Sons, Inc., Hoboken, New Jersey
Published simultaneously in Canada.

For general information on our other products and services please contact our Customer Care Department within the U.S. at 877-762-2974, outside the U.S. at 317-572-3993 or fax 317-572-4002.

Wiley also publishes its books in a variety of electronic formats. Some content that appears in print, however, may not be available in electronic format.

Library of Congress Cataloging-in-Publication Data is available.

ISBN-13 978-0-470-09733-5
ISBN-10 0-470-09733-7

10 9 8 7 6 5 4 3 2 1

Contents

Modelling

Testing

Dedication

This volume is dedicated to the memory of Geoff Frohnsdorff for his vision and leadership in the area of the materials science of construction materials. He will be best remembered as the first scientist who promoted and pursued the importance of using computer modeling as a tool in understanding cement hydration, and a tireless advocate for the use of materials science in concrete performance, including most especially durability and service life prediction.

Born in London, Dr. Frohnsdorff served two years in the Royal Air Force. He then began studies at the University of St. Andrews, receiving bachelor's degrees in mathematics and chemistry in 1951, natural philosophy in 1952 and chemistry in 1953. A Fulbright travel grant first brought him to the United States, where he received a master's degree in physical chemistry from Lehigh University in 1956. He returned to the United Kingdom in 1955, receiving a doctorate and a second advanced degree in physical chemistry from the University of London's Imperial College of Science, Technology and Medicine in 1959. He spent a year doing postdoctoral research at the Royal Military College of Canada. He then worked for 10 years at the American Cement Corp. in Riverside, Calif., at which he led a group of scientists in development of the first computer model of cement hydration. Geoff moved to Gaithersburg, Maryland in 1970, the year in which he also became a U.S. citizen, and began working for the Gillette Research Institute in Rockville. In 1973, he began his career with NIST that was to bring him national and international recognition.

During an almost a 30 year career at the Department of Commerce's National Institute of Standards and Technology (NIST, formerly NBS), he established and developed a building materials program that improved standards, test methods, computer modeling, and cement and concrete technology worldwide. As chief of NIST's Building Materials Division, Dr. Geoff Frohnsdorff oversaw research on the durability of building materials and the operation of the Construction Materials Reference Laboratory until he retired in 2002. It was his idea to establish the Cements Division of The American Ceramic Society. During his career, he was awarded the Bryant Mather award, the Bates Memorial award, and the Cavanaugh award from ASTM, in which he held the rank of Fellow. He was also a Fellow of The American Ceramic Society and American Concrete Institute and a former member of the ACI Board. He received one of the Department of Commerce's highest awards, the Silver Medal.

Despite his serious medical condition at the time, he traveled 24 hours non-stop from the UK to attend this workshop, which happened to be his last technical meeting. Geoff delivered an informal overview talk during the closing session, and it was shortly after the workshop that he was diagnosed with terminal cancer.

Geoff will be remembered for being among the pioneers of our field. In the last few years of his life, Geoff said that his career had consisted of "two or three good ideas, and lots of signing of bits of paper." When asked what these ideas were, he said that they were service life prediction, the computer modeling of concrete, and the use of materials science in concrete technology and standards. This short testimonial to Geoff could be thought of as his "signing" of these proceedings (another bit of paper?!). We want to pay tribute to his personality, charm, vision, determination, and most of all, his "two or three good ideas," which have greatly influenced the direction of the field of the materials science of concrete.

Preface

We have a staggering societal demand for the utilization of cement based materials. Concrete production in the US has almost doubled from 220 million cubic yards per year in early 1990's to more than 430 million cubic yards in 2004, while the worldwide production and use of concrete will soon surpass the 10 billion tons per year mark. From a technical perspective, numerous challenges remain in order to manufacture, promote, and use concrete materials in a sustainable manner. These challenges deal with an integration of various aspects of cement chemistry, transport properties, micromechanics, basic engineering guidelines, and industry specifications. By predicting and controlling microstructural changes and chemical reactions during the life cycle of products we can better address damage evolution, and premature degradation of infrastructure systems. This would allow us to better utilize our resources.

From a sustainability point of view, we must address material shortages along with initial and life cycle maintenance costs of construction projects. Other aspects of specifications and quality control are also directly related to how we understand, measure, and communicate with measures of quality and durability. These aspects must be balanced with new research findings which enable opportunities for new knowledge, applied to new and old products to address the ever-expanding needs of our society. The idea of this workshop was formed based on a need to better integrate aspects of materials science, mechanics, modeling, and testing in developing tools of understanding the durability in cement based materials. The goals were based on generating cross disciplinary tools to guide us toward a more sustainable, intelligent, and responsible manufacturing and engineering policy.

One of the basic issues in deterioration of concrete structures is related to transport of fluids through the cementitious matrix. A workshop on Transport Properties

& Concrete Quality was held at the campus of Arizona State University on October 10-12, 2005. During this meeting, scientific aspects of the relevant relationships between materials, mechanisms, processes, and service life were discussed. Areas covered at the workshop included field observations, fundamentals, and applied state-of-the-art all communicated in terms of the relationship between transport properties, durability, and quality of concrete infrastructure. Approaches that combined different techniques, disciplines, and length-scales were combined to study the inter-relationships among permeation, diffusion, microstructure, pore structure, and transport mechanisms. Other aspects discussed during the meeting involved hydration mechanisms and role of supplementary materials, deterioration, mathematical and computer modeling, testing, specifications, and field experiences.

The program consisted of nine technical sessions ranging from the current state of practice, to theory, simulation, and testing and specifications. The collection of papers in this volume represents the technical presentations held during the workshop.

We wish to express our most sincere gratitude to the sponsors of this workshop who through their generous contribution made this event possible. The sponsors include: ASU Pavements and Materials Conference Committee, Lafarge Corporation, Grace Construction Products, Salt River Project, NIST, Arizona Cement Association, American Coal Ash Association, Degussa Admixtures, Forta Corporation, and MACTEC Engineering & Consulting, Inc. Our gratitude is extended to The American Ceramic Society for their patience and support with the publication of the proceedings.

We also acknowledge the support and encouragement of Professor Sandra Houston, Chair of the Civil and Environmental Engineering Department at ASU. Ms. Dawn Takeuchi, and Mr. Peter Goguen of CEE department graciously helped with many aspects of the workshop. The efforts of Ms. Judy Reedy for her tireless work in organization, planning, and coordination prior to and during the workshop deserves a sincere and special note of appreciation.

BARZIN MOBASHER AND JAN SKALNY

Mechanisms

TEN OBSERVATIONS FROM EXPERIMENTS TO QUANTIFY WATER MOVEMENT AND POROSITY PERCOLATION IN HYDRATING CEMENT PASTES

Dale P. Bentz
Building and Fire Research Laboratory
National Institute of Standards and Technology
100 Bureau Drive Stop 8615
Gaithersburg, MD 20899-8615

ABSTRACT

The transport properties and durability performance of concrete structures are both strongly influenced by the three-dimensional microstructure that is established during early age placement and curing. This paper will present observations from two experimental techniques that have been applied to examining this early age microstructure development. First, x-ray absorption measurements are applied to study water (and cement particle) movement during settlement, drying, and curing. Five observations from a series of experiments conducted on single layer and bilayer composite specimens are presented and supported by experimental data. The influences of curing conditions, water-to-cement mass ratio (w/c), cement particle size distribution, shrinkage-reducing admixtures (SRAs), and internal curing via the addition of saturated lightweight aggregates (LWAs) are highlighted. Second, low temperature calorimetry (LTC) is utilized to examine the depercolation/repercolation of the capillary pores in a hydrating cement paste as a function of w/c, curing temperature, and the addition of various alkali ions or an SRA. Once again, five observations from this second type of experiment are presented and substantiated by experimental data. These ten observations have numerous implications for concrete curing practices, possible new applications for existing admixtures, and durability performance of field concrete.

INTRODUCTION

It is well known that the microstructure (particularly the pore structure) of concrete largely determines its transport properties and in many cases it ultimate service life. The pore structure is a dynamic system that changes dramatically during placement, curing and hydration, and field exposure and aging. Additionally, pores can be filled with water, vapor, or some combination of the two and the medium filling the pores will have a large influence on continuing hydration, transport, and common degradation reactions (carbonation for example). This paper focuses on two aspects of this developing pore structure, namely water movement within and through a cement paste microstructure during early age curing/drying and depercolation/repercolation of the capillary pores due to hydration and aging. The former process is investigated via the application of x-ray absorption measurements, while the latter is studied using low temperature calorimetry (LTC) experiments. Results are conveniently summarized as a set of ten critical observations of relevance both to basic research and to field applications.

EXPERIMENTAL METHODS

X-ray Absorption

The x-ray absorption technique and measurement procedures have been described in detail in a variety of publications.[1-5] The basic procedure is to expose the specimen of interest to a concentrated beam of x-rays and monitor the "quantity" of x-rays that are transmitted through the specimen. Previously, a point detector based on either a NaI or a Cd-Zn-Te (CZT) crystal has been employed,[1-5] but in the latest version of the commercially available equipment, a 256 x 256 camera detector with an active area of 25 mm x 25 mm is utilized. Specimens are placed in fixed locations between the x-ray source and the detector and sampled in both the spatial and temporal domains. Generally, the results are recorded as the number of counts received by the detector during a fixed integration time, normalized by the counts detected through a reference material specimen (such as water or dry cement powder) in the same integration time. In the graphs presented below, these count values are plotted against the spatial location within the specimen (thickness) for responses obtained at various curing times. More counts indicate a higher transmittance of the incoming x-rays and are generally interpreted as corresponding to a less dense material. For hydrating (and drying) cement paste, an increase in counts could indicate either a replacement of cement particles by water (during settling and bleeding for example) or a decrease in water content (after set is achieved) due to either drying or local rearrangement of water due to chemical shrinkage and self-desiccation. Previously, the relative standard uncertainty in the normalized counts has been determined to be on the order of 0.3 %.[2]

Low Temperature Calorimetry

The low temperature calorimetry technique has also been described in several recent publications.[6-8] A commercially available differential scanning calorimeter (DSC) with a cooling unit was utilized to cool small specimens of cement paste (mass typically between 30 mg to 90 mg) from 5 °C to -55 °C at a controlled rate of -0.5 °C/min. For temperatures between -100 °C and 500 °C, the DSC manufacturer has specified a constant calorimetric sensitivity of ± 2.5 %, with a root-mean-square baseline noise of 1.5 μW. Typical measured signals for a freezing scan are on the order of 0.5 mW to 1 mW. The peaks observed in a plot of heat flow versus temperature correspond to water freezing in pores with various size entryways (pore necks). The smaller the pore entryway, the more the freezing peak is depressed. Thus, a larger isolated water-filled pore will not freeze until the water in the smaller entryway pores surrounding it first freezes. The presence of, absence of, or change in these peaks can be used to infer critical information concerning the characteristic sizes of the "percolated" (connected) water-filled pores in the microstructure of the hydrating cement pastes. One advantage of LTC over mercury intrusion porosimetry and other techniques for assessing pore size and connectivity is that the specimens may be evaluated without any applied drying that might damage the pore structure.

OBSERVATIONS

1) Microstructural gradients established during settling/drying are different under sealed (saturated) and exposed (drying) conditions.

After casting, the microstructure of fresh cement paste can undergo significant changes due not only to settling (gravity) and bleeding, but also due to drying. For cement paste specimens 5 mm to 30 mm thick, the microstructural gradients developed through their thickness depend strongly on their w/c and the exposure conditions. As shown in Figure 1, for sealed curing, the initial settling may result in a densification of the cement paste in the lower portion of the specimen, as indicated by the smaller number of x-rays (counts) that are transmitted through the width of the specimen in the bottom portion (31,000) vs. the top portion (34,000). As cement particles descend and water "rises" during settlement and bleeding, the lower portions of the specimens are naturally densified (fewer x-rays transmitted) more than the upper portions. Thus, due to these settlement processes, under sealed curing conditions or when water is ponded on top of the specimens, it would be expected that the upper portions of the specimens (surface layer) would exhibit a higher local w/c and ultimately a more porous microstructure. As illustrated in Figure 2, much less microstructure rearrangement of this type was observed in a w/c=0.3 cement paste, as in a denser, more viscous paste, settlement and its accompanying bleeding are nearly nonexistent, and a basically flat x-ray profile is observed.

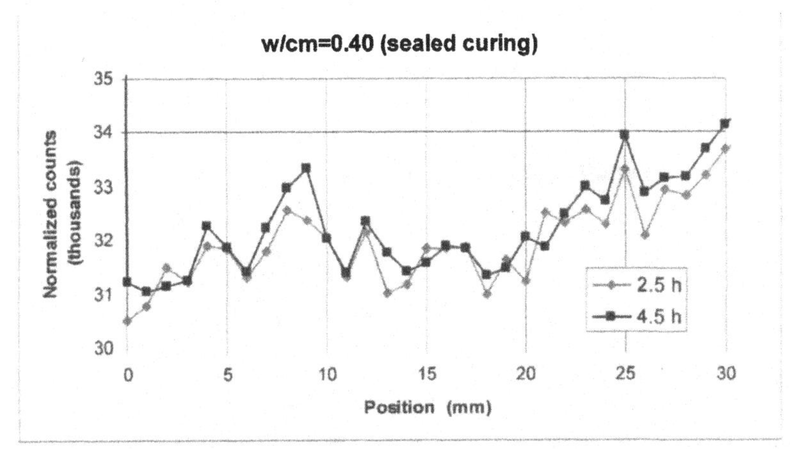

Figure 1. Normalized counts for x-rays transmitted through a w/cm=0.40 blended cement paste cured under sealed conditions at 23 °C, as a function of curing time (2.5 h or 4.5 h) and depth.[4] Top of specimen is located at 30 mm.

When drying is superimposed on the settlement/bleeding process, additional capillary forces may be imposed at the drying surface, due to the menisci that are formed there as the drying front begins its initial penetration into the porous (set) microstructure.[9,10] These forces will induce an extra densification at the top (exposed) surface, so that in this case, the microstructural gradient as a function of depth may remain relatively flat or may even exhibit a preferential densification near the top surface, as exemplified in Figure 3,[1] particularly focusing on the differences between the 0.67 h and the 4.67 h scans. From this perspective, the recommended practice of applying a curing compound only when the top surface of the concrete first appears "dry" and free of surface water[11] should also be beneficial in promoting the formation of an equal or superior quality surface layer in the concrete.

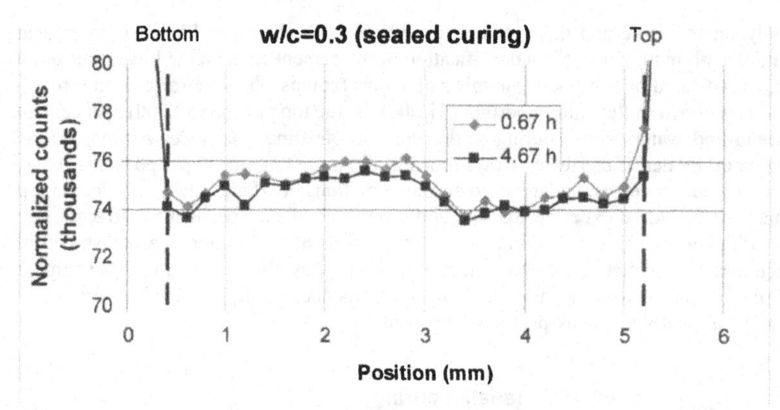

Figure 2. Normalized counts for x-rays transmitted through a w/c=0.30 cement paste cured under sealed conditions at 23 °C, as a function of curing time (0.67 h or 4.67 h) and depth.[1]

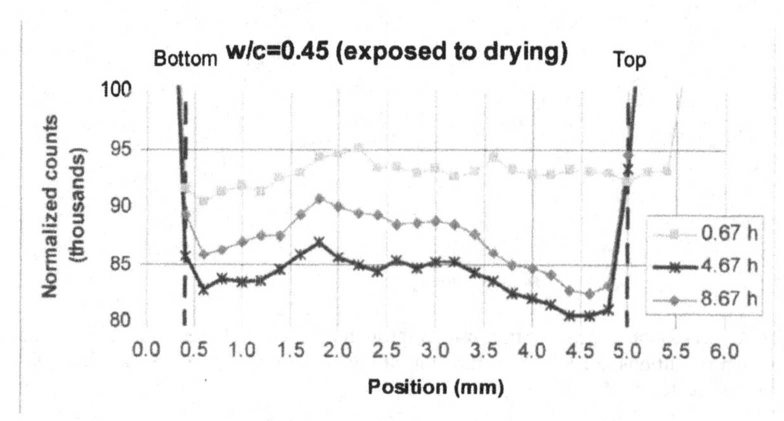

Figure 3. Normalized counts for x-rays transmitted through a w/c=0.45 portland cement paste immediately exposed to drying conditions (23 °C and 50 % RH), as a function of curing time (0.67 h, 4.67 h or 8.67 h) and depth.[1] Top of specimen is located at about 5 mm.

2) After settling, 10 mm to 20 mm thick ordinary portland cement paste specimens without admixtures and with w/c $\leq$ 0.45, dry out "uniformly" through their thickness.

Because the cement particles in a typical cement paste have a fairly wide particle size distribution (diameters ranging from submicron to about 100 μm), the pore size distribution in the fresh paste (dispersion) also exhibits a wide range of pore sizes. Thus, unlike many materials which exhibit a sharp drying front that "intrudes" from the exposed surface inward, after settling, ordinary portland cement pastes with w/c $\leq$ 0.45 dry out rather uniformly, as illustrated by the results shown in Figure 4. This "uniform" drying implies that the largest pores (some of which could be part of bleed channels) throughout the thickness of the specimen are the first to empty when the specimen is exposed to a drying environment. It should be noted that results similar to those shown in Figure 4 were obtained for a w/c=0.3 cement paste where observable bleeding was minimal.[1] Similar results have also been obtained for a much thicker (50 mm) cement paste specimen using magnetic resonance imaging.[12] The potential implications of this observation for curing of field concrete are critical, as the drying that is observed to be occurring at the exposed concrete surface can actually be influencing the water content and microstructure of the hydrating cement paste at much greater depths, perhaps even at the depth of the steel reinforcement.[1]

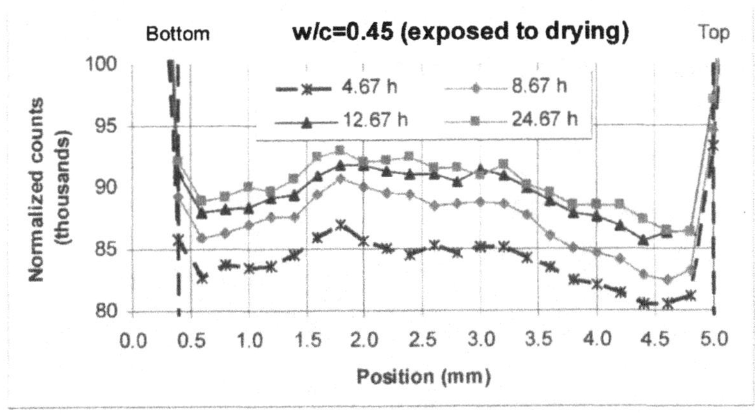

Figure 4. Normalized counts for x-rays transmitted through a w/c=0.45 portland cement paste immediately exposed to drying conditions (23 °C and 50 % RH), as a function of curing time (4.67 h, 8.67 h, 12.67 h, or 24.67 h) and depth.[1] Top of specimen is located at about 5 mm.

3) In bilayer composite specimens, water always first moves preferentially from a coarser pore structure to a finer one during either drying or sealed hydration. Eventually, during a drying exposure, the finer layer will lose water as well. A composite specimen with a finer pore structure layer on top of a coarser one will lose less water than one with the two layers reversed, under equivalent drying exposures.

During the initial experiments using the x-ray absorption technique, a variety of bilayer composite cement paste specimens were prepared.[1,2] The two parameters varied between the two layers were w/c (for a fixed cement powder) and cement particle size distribution (for a fixed w/c of 0.45). In the former case, a higher w/c paste will have a coarser pore size distribution due to its higher water volume fraction, while in the latter case, the coarser cement will produce a coarser pore size distribution due to the increased spacing between the (fewer) cement particles. As illustrated in Figures 5 and 6, regardless of how the coarser pore structure is produced, water always first moves preferentially from the coarse pore structure layer to the finer one, even when it is the finer one that is directly exposed to the drying environment. In Figure 5, with a lower w/c cement paste layer over a higher w/c one, the higher w/c layer is seen to lose water during the first 7 h while the exposed lower w/c layer on top remains basically saturated, even imbibing a small amount of additional water to offset that "consumed" by chemical shrinkage during hydration. For 11 h and beyond, further water loss from the higher w/c layer is minimal and the lower w/c layer also begins to dry out. In Figure 6, with a cement paste layer produced using a finer cement (654 m^2/kg) over one produced using a coarser cement (254 m^2/kg), the coarser layer is seen to lose water during the first 5 h while the finer layer remains saturated. For 7 h and beyond, further water loss from the coarser layer is minimal and the finer layer also begins to dry out. In both Figure 5 and Figure 6, each of the 4 mm to 5 mm thick cement paste layers is observed to dry out in a relatively uniform fashion throughout their thickness once drying begins, in agreement with observation # 2. It should be further noted that the initial (1 h or 3 h) microstructural gradients established through the thicknesses of the various layers are consistent with observation # 1 above.

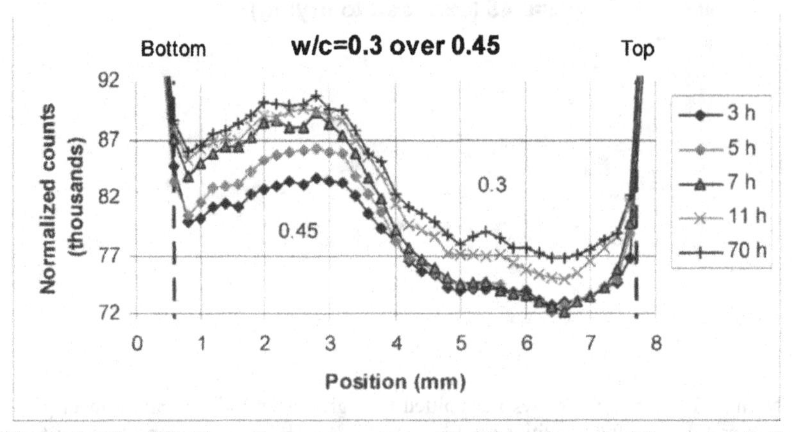

Figure 5. Normalized counts for x-rays transmitted through a w/c=0.3 over a w/c=0.45 bilayer portland cement paste immediately exposed to drying conditions (23 °C and 50 % RH), as a function of curing time (3 h to 70 h) and depth.[1] Top of specimen is located at about 7.5 mm and the lower w/c cement paste layer is directly exposed to the drying environment.

This preferential water movement from a coarser pore structure to a finer one has implications both for curing and for the application of repair materials. In the former case, controlled permeability formwork[13] and internal curing[14] are two examples of how these principles can be used to advantage. In the latter case, the saturation level of the material being repaired and the pore structure of the repair material must be carefully coordinated to avoid significant water movement from the repair material to the in place material, while perhaps promoting some water movement from the in place material to the repair material to enhance its hydration and offset any water loss to the external environment.

4) Shrinkage-reducing admixtures (SRAs) significantly modify the drying profile of exposed cement paste specimens and reduce their drying rates. In a bilayer composite, water is preferentially first drawn from a layer with SRA into a layer without SRA.

Because SRAs significantly reduce the surface tension of the cement paste pore solution, they can have a large impact on both the drying rate and the shape of the drying profile.[3,9,10] As illustrated in Figure 7, when a 2 % SRA addition by mass of cement is made to a w/c=0.35

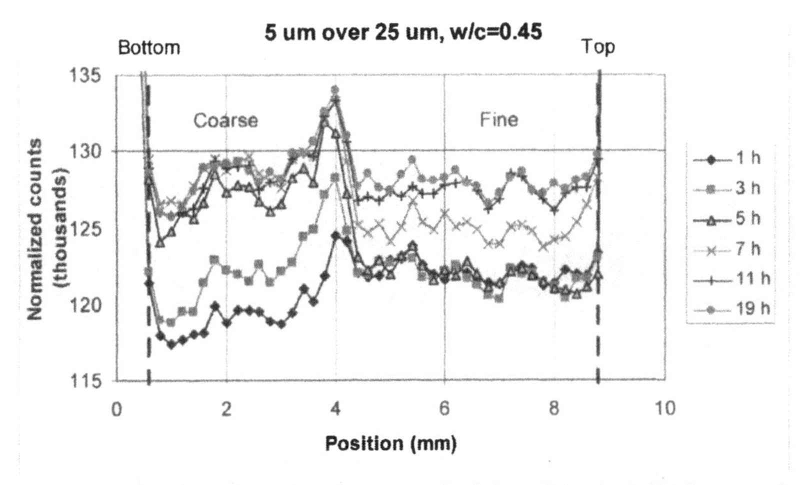

Figure 6. Normalized counts for x-rays transmitted through a w/c=0.45 bilayer portland cement paste immediately exposed to drying conditions (23 °C and 50 % RH), as a function of curing time (1 h to 19 h) and depth.[2] Top of specimen is located at about 9 mm and the finer cement paste layer is directly exposed to the drying environment.

cement paste, a sharp(er) drying front is formed intruding downward from the exposed surface, in contrast to the "flat" profile observed in Figure 4, for example. The drying front maintains its form for about the first 6 h of drying, followed by more uniform drying throughout the specimen thickness. It has been proposed[3] that the initial water removal from the exposed surface concentrates the SRA present in that top layer. As this concurrently decreases the surface tension of the pore solution in the top layer, that layer is unable to draw the higher surface tension pore solution up from deeper within the specimen. Hence, the initial penetration of the drying front into the specimen is observed. This initial penetration of the drying front also substantially slows the drying process as the porous material containing the SRA enters its "capillary regime" of drying, while a material without the SRA remains in its higher drying rate "evaporative regime".[3,9,10,15] Two practical implications of this observation have recently been indicated, namely the usage of SRAs to reduce plastic shrinkage cracking[10] and the application of a curing solution containing 10 % to 20 % SRA to reduce evaporative water loss and thus enhance hydration.[9]

5) During hydration, in systems with internal curing, water movement from water reservoirs to the surrounding hydrating cement paste can be detected using x-ray absorption measurements.

Previously, the x-ray absorption system with a point detector has been utilized to monitor water movement from a saturated lightweight aggregate (LWA) to adjacent hydrating cement paste during early age curing.[5] The amount of water movement indicated by the change in x-ray transmittance correlated linearly with the chemical shrinkage of the hydrating cement paste, as measured in a separate experiment or as modeled using the CEMHYD3D hydration and microstructure development software.[5] More recently, similar x-ray transmittance data have

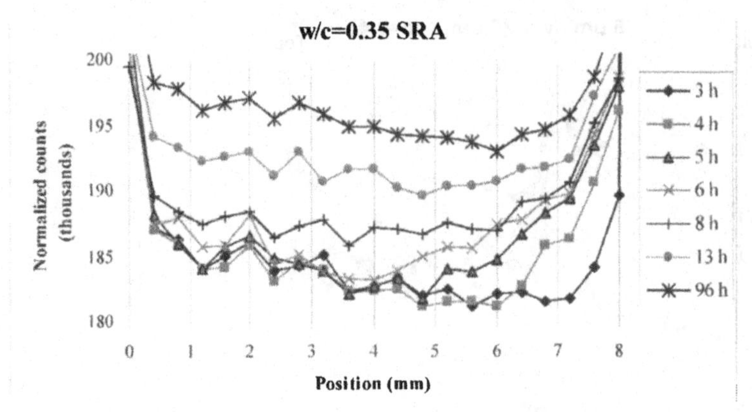

Figure 7. Normalized counts for x-rays transmitted through a w/c=0.35 portland cement paste with a 2 % SRA addition by mass of cement, immediately exposed to drying conditions (23 °C and 50 % RH), as a function of curing time (3 h to 96 h) and depth.[3] Top of specimen is located at about 8 mm.

been obtained utilizing the new camera detector that provides a complete 256 x 256 pixel image of the materials being investigated, with a resolution of 0.1 mm/pixel. A single ≈10 mm saturated LWA particle was surrounded by a w/c=0.3 cement paste and placed in a thin sealed plastic container (width of about 8 mm). A rubber gasket was used as an exterior guard ring surrounding the cement paste/LWA specimen. The container was placed between the x-ray source and the camera detector and x-ray images were acquired after various hydration times, as illustrated in Figure 8. The images obtained by subtracting the image at 0.5 h from those obtained after different hydration times clearly indicate the preferential movement of water out of the saturated LWA into the surrounding cement paste to satisfy the water demand in the cement paste, created by the chemical shrinkage that is accompanying the ongoing hydration. In the future, it is planned to conduct three-dimensional x-ray microtomography experiments at a higher spatial resolution on mortars with internal curing, to hopefully observe this same process in 3-D.

6) In low w/c cement pastes hydrated under sealed conditions, capillary porosity first depercolates, but later repercolates due to self-desiccation stresses and C-S-H rearrangement (and possibly microcracking).

Typical LTC scans for a w/c=0.35 cement paste hydrated at 20 °C under saturated conditions are provided in Figure 9. Three peaks are commonly observed in the LTC scans: one near -15 °C corresponding to water in percolated capillary pores, one at about -25 °C corresponding to open gel pores (entryways), and one between -40 °C and -45 °C, corresponding to dense gel pores, using the naming convention proposed by Snyder and Bentz.[6] The scans in Figure 9 follow a typical evolution as the capillary pores become depercolated somewhere between 3 d and 4 d, followed by a later depercolation of open gel pores between 14 d and 30 d. Beyond this time, the only peak observed in the LTC scans corresponds to those pores that are percolated via entryways composed of the dense gel pores, and the height of this peak is

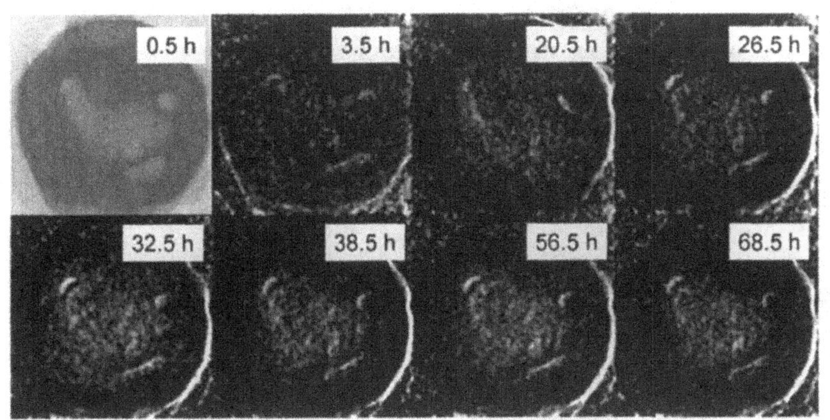

Figure 8. X-ray images for internal curing of w/c=0.3 hydrating cement paste surrounding a single initially saturated LWA particle. Top left image is initial x-ray view of system with brighter (less dense) saturated LWA particle as central portion of circular specimen, surrounded by a rubber gasket. Subsequent images are subtractions of this initial image from images obtained after various hydration times. In the subtracted images, brighter areas indicate drying while darker areas indicate wetting. Hydration times are as indicated. The slight halo observed at the right specimen edge indicates a small misalignment of the specimen during movement of the x-ray source and camera detector between consecutive images.

continually decreasing along with the total porosity of the specimen due to continuing hydration. The height of the capillary pore peak near -15 °C has recently been successfully correlated to the connected (percolated) fraction of capillary porosity as predicted by the CEMHYD3D model.[16]

Even more interesting results are obtained when sealed curing is considered. Figure 10 presents a series of LTC scans for w/c=0.35 specimens cured under sealed conditions and then resaturated for a single day prior to the LTC scan. Similar to the observations made previously by Bager and Sellevold concerning the influence of drying/resaturation on the LTC scans,[17] it is observed that during sealed curing, while the capillary pores do initially depercolate, they later (14 d and beyond) repercolate due to an "aging" process that likely includes self-desiccation, autogenous strain development, creep, and possibly microcracking. This repercolation is even observed in very low w/c=0.25 cement pastes cured under supposedly "saturated" conditions, as the initial depercolation of the capillary porosity reduces the rate of water imbibition below that needed to maintain saturated conditions within these small (2 mm to 5 mm thick) specimens.[16] As observed in Figure 10, the measured "damage" becomes more severe with age; using the correlation developed via the CEMHYD3D model, the observed peak heights in Figure 10 can be conveniently converted to a damaged or repercolated pore volume fraction.[16]

7) This repercolation has also been observed in a low alkali w/c=0.40 cement paste hydrated under saturated curing conditions. Alkali additions, particularly lithium, to this paste appear to "stabilize" the C-S-H so that this repercolation was not observed.

It has been observed that the alkali level of the portland cement also has a strong influence on porosity depercolation/repercolation, likely via its influence on the morphology and

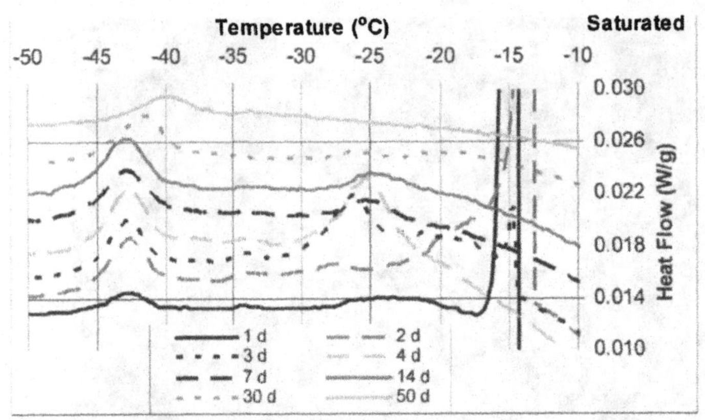

Figure 9. LTC scans for a w/c=0.35 ordinary portland cement paste cured under saturated conditions at 20 °C for various ages.

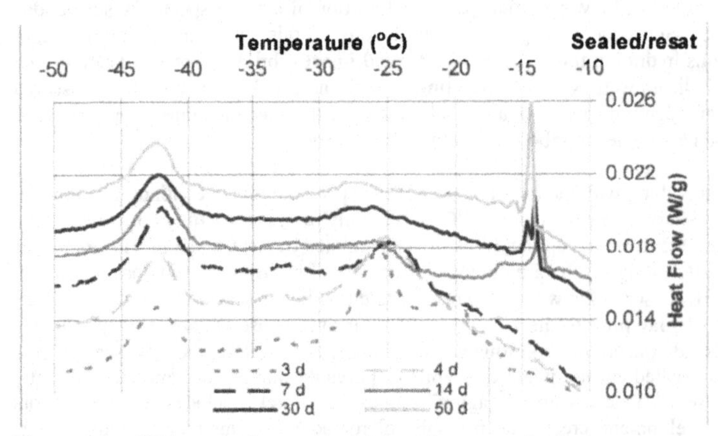

Figure 10. LTC scans for a w/c=0.35 ordinary portland cement paste cured under sealed conditions at 20 °C for various ages and then resaturated for one day.

crystallinity of the calcium silicate hydrate (C-S-H) gel that is the dominant reaction product of portland cement hydration. For example, as shown in Figure 11, for a low alkali w/c=0.4 cement paste (0.093 % Na_2O and 0.186 % K_2O per unit mass of cement), saturated curing for an extended period of time (over 100 d) results in the (re)formation of a highly percolated capillary pore structure in a system where the capillary porosity had initially depercolated due to hydration.[8] For a w/c=0.4 cement paste, it is unlikely that the observed repercolation is due to the "self-desiccation" aging described above. Instead, it appears that the C-S-H in the low alkali cement paste is more freely able to undergo rearrangement (some combination of shrinkage and creep) even in a saturated environment. As illustrated in Figure 11, the addition of extra alkali ions seems to stabilize the C-S-H so that this repercolation is largely not observed.[8] Several researchers have indicated that the presence of increased alkalis tends to produce a more crystalline (stable) C-S-H with a greater propensity towards a plate (lath)-like structural morphology.[18,19] According to recent three-dimensional microstructure-based simulations, plate-like C-S-H microstructures would be expected to be more efficient at depercolating an originally connected capillary pore structure than ones based on a totally random morphology.[20]

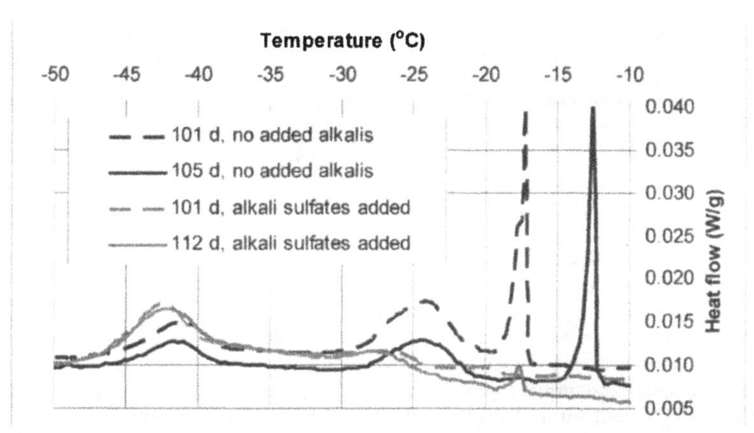

Figure 11. LTC scans for a w/c=0.40 low alkali portland cement paste cured under saturated conditions at 20 °C, with and without the addition of alkali sulfates.[7,8]

8) While hydration is indeed accelerated at higher temperatures (e.g., 40 °C), it takes more time (and hydration) for the capillary pores to depercolate, implying a "coarser" pore structure for high temperature curing.

It is well known that an increase in temperature accelerates hydration rates in cement-based materials and generally produces a coarser pore structure.[21,22] Thus, it may come as no surprise, that as shown in Figure 12, for hydration at 40 °C, the depercolation of the capillary porosity in a w/c=0.35 cement paste is delayed to occur between 7 d and 14 d of saturated curing, as opposed to between 3 d and 4 d observed for the 20 °C curing of a similar paste shown in Figure 9. At the higher curing temperature, the depercolation thus takes longer both in terms of time and in terms of the necessary degree of hydration. This observation would be consistent with the internal relative humidity measurements as a function of curing temperature recently presented by Persson,[23] and would suggest the formation of a denser C-S-H gel at higher temperatures. The CEMHYD3D version 3.0 computer model[24] has recently been modified to incorporate this effect by having the local C-S-H precipitation be a function of curing temperature and good agreement has been observed between the revised model and the experimental data.[16] As with 20 °C curing, sealed/resaturated w/c=0.35 cement pastes also exhibit (re)percolated capillary pore structures after 28 d or more of sealed curing at 40 °C .

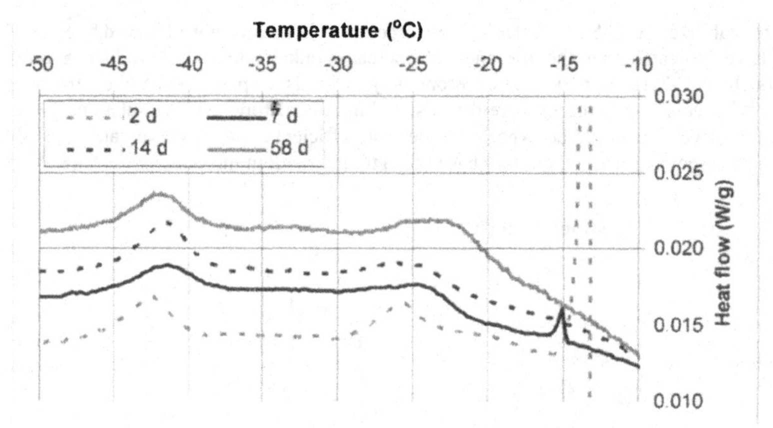

Figure 12. LTC scans for a w/c=0.35 ordinary portland cement paste cured under saturated conditions at 40 °C for various ages.

9) Sealed/saturated curing can be beneficial for an earlier depercolation of the capillary porosity in intermediate (e.g., 0.4 to 0.45) w/c cement pastes.

While much emphasis is currently placed on maintaining saturated conditions in concrete during (external and internal) curing, many years ago, Swayze[25] and Powers[26] separately advocated the possible usage of a sealed/saturated curing regimen, the former to offset the expansion and cracking due to thermal effects in large structures, and the latter to promote frost resistance in fresh concrete. As exemplified in Figure 13, for an intermediate range of w/c ratios (likely 0.40 to 0.45) an initial period of sealed curing followed by ponding of water on top of the specimen (saturated curing) results in a more depercolated pore structure after 14 d total curing time than when saturated curing conditions are maintained throughout the 14 d. For this particular w/c=0.45 cement paste, 8 d of sealed curing followed by saturation appears to be superior to a 3 d sealed/11 d saturated regimen. The initial sealed curing is more effective at depercolating the capillary porosity, because it concentrates hydration in the smaller pores and pore entryways (and not in the empty larger pores formed due to chemical shrinkage and self-desiccation).[7]

10) Because an SRA will significantly reduce the surface tension of the pore solution, it can also shift the freezing point depression in small pores to higher temperatures, so that significantly more freezable water is present at any given temperature for curing ages of up to several weeks.

Because SRAs can reduce the surface tension of water (and pore solution) by as much as 50 %,[3,10] via the Kelvin-Laplace equation, the capillary stresses in any partially water-filled pores will be reduced proportionally. Similarly, the freezing point depression in a given size pore is also a function of the surface tension, although some controversy exists as to whether it should be the liquid-gas surface tension or the solid-liquid surface tension.[27] If one were to assume that it is the liquid-gas surface tension (as these values have already been measured in the literature[3,10] and appear to give good agreement with the experimental results of Powers and Brownyard for cement paste[27]), for a 50 % reduction in surface tension due to the addition of an

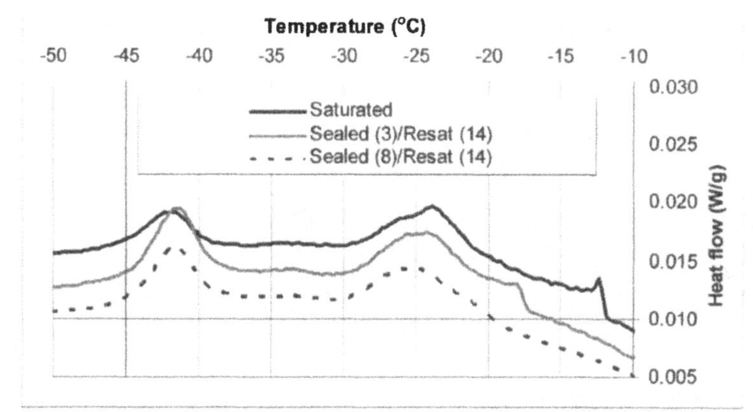

Figure 13. LTC scans for a w/c=0.45 ordinary portland cement paste cured under "sealed/saturated" conditions at 40 °C for 14 d of total curing time.

SRA, one would expect freezing point depressions of -45 °C and -25 °C to be shifted to values of about -19 °C and -12 °C, respectively. Thus, these peaks would be occurring in the same temperature range near -15 °C where freezing in the percolated capillary pores is usually observed at the scanning rates of -0.5 °C/min typically employed in these LTC experiments.

As shown by the results presented in Figure 14, this indeed seems to be the case[28] for a w/c=0.35 cement paste with a 3 % SRA addition by mass of cement cured at 20 °C under saturated conditions. At ages up to 14 d, the only peaks observed in the LTC scans are in the temperature range of about -12 °C to about -20 °C, and in some cases, two separate peaks are observed for a given specimen within this temperature range. It is only at later ages that measurable peaks appear near -25 °C (14 d and 21 d) and -45 °C (21 d) for the open and dense gel pores usually detected under saturated curing (after 1 d to 2 d in Figure 9). This could suggest that this particular SRA is being effectively removed from the "free" pore solution (specifically in the smaller pore entryways) after 21 d of hydration at 20 °C. Further support for these hypotheses can be found in Figure 15, where the LTC scans for specimens cured between 14 d and 21 d under various combinations of saturated and sealed/resaturated curing are presented. A shift of the LTC peaks to lower temperatures, that would be consistent with a removal of the SRA from the pore solution in the smaller pore entryways during hydration, is clearly indicated. While the experimental results in Figures 14 and 15 are consistent with the hypotheses made above, at least two other possibilities must be recognized: the SRA could modify (coarsen) the nanostructure of the C-S-H gel such that the entryway pores sizes (formed during early age hydration) are increased substantially or it could modify the solid-liquid surface tension in the same proportion as it modifies the liquid-gas surface tension. Further experiments are underway to clarify the influence of SRA on freezable water content,[28] but the undeniable fact that can be taken from comparing Figure 14 to Figure 9 is that whatever the reason, at ages up to 14 d the cement paste prepared with the 3 % SRA addition has substantially more freezable water at temperatures in the range from -10 °C to -20 °C than the control specimen prepared without SRA. Since SRAs also sometimes function as air detrainers[29] in the fresh concrete mixture, their possible detrimental influence on early age frost resistance may be a topic worthy of further consideration.

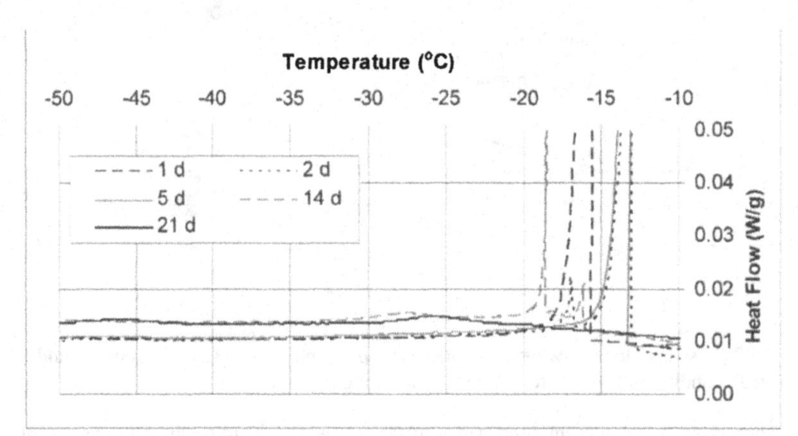

Figure 14. LTC scans for a w/c=0.35 ordinary portland cement paste with a 3 % SRA addition by mass of cement, cured under saturated conditions at 20 °C for various ages.[28]

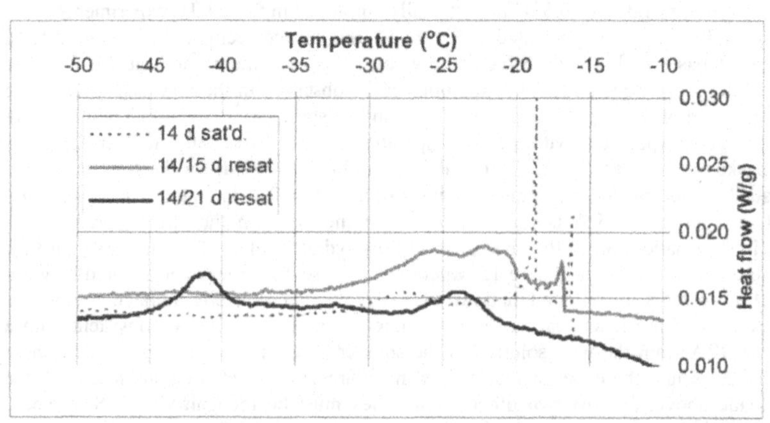

Figure 15. LTC scans for a w/c=0.35 ordinary portland cement paste with a 3 % SRA addition by mass of cement, cured at 20 °C with either 14 d of saturated curing or 14 d of sealed curing, followed by resaturation and further curing for either 1 d or 7 d.

CONCLUSIONS

X-ray absorption and low temperature calorimetry are valuable tools for studying curing and hydration of cement paste, the former for examining water movement during initial placement and curing, and the latter for quantifying the percolation state of the capillary porosity during hydration and aging. Many of the ten observations presented in this paper would not have been predicted a priori, so both techniques are clearly advancing the state of knowledge in the field of cement-based materials science research. In addition, numerous practical (field) implications of the observations were provided in the contexts of curing, transport properties, and durability of cement-based materials.

ACKNOWLEDGEMENTS

The author would like to acknowledge useful discussions with Dr. Pietro Lura of the Technical University of Denmark and thank Dr. Chiara Ferraris and Dr. Edward Garboczi of BFRL/NIST for their thorough reviews of the manuscript.

REFERENCES

[1]Bentz, D.P., and Hansen, K.K., "Preliminary Observations of Water Movement in Cement Pastes During Curing Using X-ray Absorption," *Cem. Concr. Res.*, **30**, 1157-68 (2000).

[2]Bentz, D.P., Hansen, K.K., Madsen, H.D., Vallee, F.A., and Griesel, E.J., "Drying/Hydration in Cement Pastes During Curing," *Mater. Struct.*, **34**, 557-65 (2001).

[3]Bentz, D.P., Hansen, K.K., and Geiker, M.R., "Shrinkage-Reducing Admixtures and Early Age Desiccation in Cement Pastes and Mortars," *Cem. Concr. Res.*, **31** (7), 1075-85 (2001).

[4]Bentz, D.P., "Influence of Curing Conditions on Water Loss and Hydration in Cement Pastes with and without Fly Ash Substitution," NISTIR **6886**, U.S. Department of Commerce, July 2002.

[5]Lura, P., Bentz, D.P., Lange, D.A., Kovler, K., Bentur, A., and van Breugel, K., "Measurement of Water Transport from Saturated Pumice Aggregates to Hardening Cement Paste," in Proc. Advances in Cement and Concrete IX: Volume Changes, Cracking, and Durability, Eds. D. Lange, K.L. Scrivener, and J. Marchand, Copper Mountain, CO, 2003, pp. 89-99.

[6]Snyder, K.A., and Bentz, D.P., "Suspended Hydration and Loss of Freezable Water in Cement Pastes Exposed to 90 % Relative Humidity," *Cem. Concr. Res.*, **34** (11), 2045-56 (2004).

[7]Bentz, D.P., and Stutzman, P.E., "Curing, Hydration, and Microstructure of Cement Paste," submitted to *ACI Mater. J.*, 2005.

[8]Bentz, D.P., "Lithium, Potassium, and Sodium Additions to Cement Pastes," submitted to *Adv. Cem. Res.*, 2005.

[9]Bentz, D.P., "Curing with Shrinkage-Reducing Admixtures: Beyond Drying Shrinkage Reduction," *Concr. Inter.*, **27** (10), 55-60 (2005).

[10]Lura, P., Pease, B., Mazzotta, G., Rajabipour, F., and Weiss, J., "Influence of Shrinkage-Reducing Admixtures on the Development of Plastic Shrinkage Cracks," submitted to *ACI Mater. J.*, 2005.

[11]Guide to Curing Concrete (ACI 308.R-01), *ACI Manual of Concrete Practice*, American Concrete Institute, Farmington Hills, MI, 2001.

[12]Coussot, P., private communication, 1999.

[13]Sousa-Coutinho, J., "Durable Concrete Through Skin Treatment with CPF," in: Durability of Building Materials and Components 8 (Vol. 1), M.A. Lacasse and D.J. Vanier (Eds.), National Research Council, Ottawa, Canada, pp. 453-62, 1999.

[14]Bentz, D.P., Lura, P., and Roberts, J., "Mixture Proportioning for Internal Curing," *Concr. Inter.*, **27** (2), 35-40 (2005).

[15]Coussot, P., "Scaling Approach of the Convective Drying of a Porous Medium," *European Phys. J. B*, **15**, 557-66 (2000).

[16]Bentz, D.P., "Capillary Porosity Depercolation/Repercolation in Hydrating Cement Pastes via Low Temperature Calorimetry Measurements and CEMHYD3D Modeling", submitted to *J. Amer. Ceram. Soc.*, 2005.

[17]Bager, D.H., and Sellevold, E.J., "Ice Formation in Hardened Cement Paste, Part II-Drying and Resaturation on Room Temperature Cured Pastes," *Cem. Concr. Res.*, **16**, 835-44 (1986).

[18]Mori, H., Sudoh, G., Minegishi, K., and Ohta, T. "Some Properties of C-S-H Gel Formed by C_3S Hydration in the Presence of Alkali," in Proceedings of the 6th International Congress on the Chemistry of Cement, Moscow, V. 2, Supplementary Paper, Section 2, 1974, pp. 2-12.

[19]Richardson, I.G., "Tobermorite/Jennite- and Tobermorite/Calcium Hydroxide-Based Models for the Structure of C-S-H: Applicability to Hardened Pastes of Tricalcium Silicate, β-Dicalcium Silicate, Portland Cement, and Blends of Portland Cement with Blast-Furnace Slag, Metakaolin, or Silica Fume," *Cem. Concr. Res.*, **34**, 1733-77 (2004).

[20]Bentz, D.P., "Influence of Alkalis on Porosity Percolation in Hydrating Cement Pastes," submitted to *Cem. Concr. Comp.*, 2005.

[21]Bentur, A., Berger, R.L., Kung, J.H., Milestone, N.B., and Young, J.F., "Structural Properties of Calcium Silicate Pastes: II. Effect of Curing Temperature," *J. Amer. Ceram. Soc.*, **62** (7-8), 362-66 (1979).

[22]Cao, Y., and Detwiler, R.J., "Backscattered Electron Imaging of Cement Pastes Cured at Elevated Temperatures," *Cem. Concr. Res.*, **25** (3), 627-38 (1995).

[23]Persson, B., "On The Temperature Effect on Self-Desiccation of Concrete," in Proceedings of the 4th International Research Seminar on Self-Desiccation and Its Importance in Concrete Technology, Eds. B. Persson, D. Bentz, and L.-O. Nilsson, Lund University, 2005, pp. 95-124.

[24]Bentz, D.P., "CEMHYD3D: A Three-Dimensional Cement Hydration and Microstructure Development Modeling Package. Version 3.0," NISTIR **7232**, U.S. Department of Commerce, June 2005, available at ftp://ftp.nist.gov/pub/bfrl/bentz/CEMHYD3D/version30.

[25]Swayze, M.A., "Early Concrete Volume Changes and Their Control," *J. Amer. Concr. Inst.*, **13** (5), 425-40 (1942).

[26]Powers, T.C., "A Discussion of Cement Hydration in Relation to the Curing of Concrete," *Proc. Highway Res. Board*, **27**, 178-88 (1947).

[27]Fagerlund, G., "Determination of Pore-Size Distribution from Freezing-Point Depression," *Matér. Construct.*, **6** (33), 215-25 (1973).

[28]Bentz, D.P., "Influence of Shrinkage-Reducing Admixtures on Freezable Water Content of Hydrating Cement Paste," to be submitted to *Cem. Concr. Comp.*, 2005.

[29]Bentz, D.P., "Capitalizing on Self-Desiccation for Autogenous Distribution of Chemical Admixtures in Concrete," Proceedings of the 4th International Research Seminar on Self-Desiccation and Its Importance in Concrete Technology, Eds. B. Persson, D. Bentz, and L.-O. Nilsson, Lund University, 2005, pp. 189-96.

ALKALI SILICA REACTIVITY OF SILICA FUME AGGLOMERATES

Maria C. Garci Juenger
Department of Civil, Architectural and Environmental Engineering
The University of Texas at Austin
1 University Station C 1748
Austin, TX 78712

Andrew J. Maas
Thornton-Tomasetti Group
12750 Merit Drive, Suite 750, LB-7
Dallas, TX 75251

Jason H. Ideker
Department of Civil, Architectural and Environmental Engineering
The University of Texas at Austin
1 University Station C 1748
Austin, TX 78712

ABSTRACT

Silica fume is widely known to reduce expansion due to alkali silica reaction (ASR). There has been recent concern, however, that agglomerated masses of silica fume particles present in sources that have been densified or pelletized may actually behave as reactive aggregates in high alkali environments and increase expansion. The study presented here explores silica fume with various agglomerate sizes from four sources using accelerated testing for ASR. It was observed that silica fume agglomerates are not necessarily alkali silica reactive, but that keeping the maximum agglomerate size to less than 100 μm appears to prevent expansion due to ASR in the silica fume.

INTRODUCTION

Alkali-silica reaction (ASR) is an expansive chemical reaction that occurs in concrete when three factors are present: 1) alkalis, originating in the cement or from other sources, 2) reactive siliceous aggregates, and 3) moisture. Hydroxyl ions attack the siliceous structure of aggregates and cause the formation of an alkali-silica gel that imbibes water and expands, causing cracking in the concrete. It has been hypothesized that agglomerates of silica fume can act as reactive siliceous aggregates, thus directly causing ASR-related expansion in concrete.[1] These agglomerates have been observed in field and laboratory concrete samples due to inadequate mixing procedures.[2,3] The alkali-silica reactivity of these agglomerates is unclear, however. Some studies have shown that the agglomerates cause expansion,[4,5] while others show that agglomerates decrease expansion similar to finely divided silica fume.[3,6,7] The literature has been summarized by the authors previously.[8,9]

The study described in this paper was undertaken to investigate the inconsistencies in the literature with regard to this subject. It was hypothesized that the inconsistencies could be due to differences in testing regimes, differences in sample preparation, and differences in silica fume sources and sizes. In the present study, the first two variables were standardized, while the last was varied.

MATERIALS AND METHODS

Silica fumes from three sources were tested in this study. Results from a fourth silica fume tested as part of a previous study are also included.[8] An undensified silica fume which contained no agglomerates was tested; a sieve analysis showed that all particles passed through an ASTM No. 100 sieve (< 100 μm). This silica fume is labeled U for undensified. A commercially available densified silica fume was tested that had 37% of material retained on the No. 100 sieve and approximately 3.5% of material larger than this sieve size (100 μm). This silica fume will be called "medium" or "M". Another silica fume was tested that had larger agglomerates (called large, "L"). This silica fume was sieved such that 100% was retained on the No. 50 sieve (300 μm); all smaller agglomerates were removed. The silica fume from the previous study[8] contained many agglomerates of various sizes and was sieved into the same size distribution as the aggregates used to make mortar specimens (called old, "O"). This gradation is specified in ASTM C 1260;[10] all agglomerates were greater than the No. 100 sieve.

Silica fume was incorporated into mortars tested according to ASTM C 1260, "Standard Test Method for Potential Alkali Reactivity of Aggregates (Mortar-Bar Method)." This test monitors expansion of mortar prisms in a 1N NaOH solution at 80°C for 14 days. Silica fume was used as a cement replacement at levels of 0, 2, 4, 6 and 10%. Only the results of 10% replacement of cement with silica fume are discussed here. The older silica fume, O, was used as an aggregate replacement rather than a cement replacement. Silica fume O replaced 5% of the aggregate, which is roughly equivalent to a 10% cement addition. Non-reactive and reactive aggregates were used for testing.

RESULTS AND DISCUSSION

Results of ASTM C 1260 expansion testing with all silica fumes at 10% cement replacement levels with non-reactive and reactive aggregates are shown in Figures 1 and 2, respectively. In Figure 1, with a non-reactive aggregate, only silica fume L caused expansion. The other silica fumes were all inert. In Figure 2, silica fumes U and O reduced expansion of the reactive aggregate, while silica fumes M and L were inert.

It was expected that the undensified silica fume would reduce expansion of the reactive aggregate, but the responses of the densified silica fumes are varied. From these results it is apparent that the differences in observed behavior of silica fume agglomerates in the literature could be due to different sources or characteristics of the materials tested. Size of agglomerates alone does not predict behavior, as silica fumes L and O both have very large agglomerates, yet have opposite effects on ASR-related expansion. Therefore, the presence of large agglomerates in a field or laboratory sample does not suggest that ASR will occur.

It is clear from previous work,[4] however, that size is an important factor in determining the reactivity of agglomerates. Results of further testing with silica fumes M and L are shown in Figure 3. For these tests, silica fume M was folded gently into the mortar after completion of mixing in order to preserve the largest agglomerates. The silica fume L agglomerates in Figure 3 were taken from the material passing the No. 100 sieve (<100 μm), significantly smaller than the original material tested. The results in Figure 3 show the opposite effects for these materials as in Figure 1. Silica fume M went from being inert in Figure 1 to reactive in Figure 3 when agglomerate size was effectively increased. Silica fume L went from being reactive in Figure 1 to inert in Figure 3 when agglomerate size was decreased by sieving. It appears from these results, that for a silica fume to be reactive it must contain a portion of material greater than 100 μm. However, the corollary is not necessarily true; that is, silica fume with material greater than 100 μm is not necessarily reactive, as seen by the performance of silica fume O.

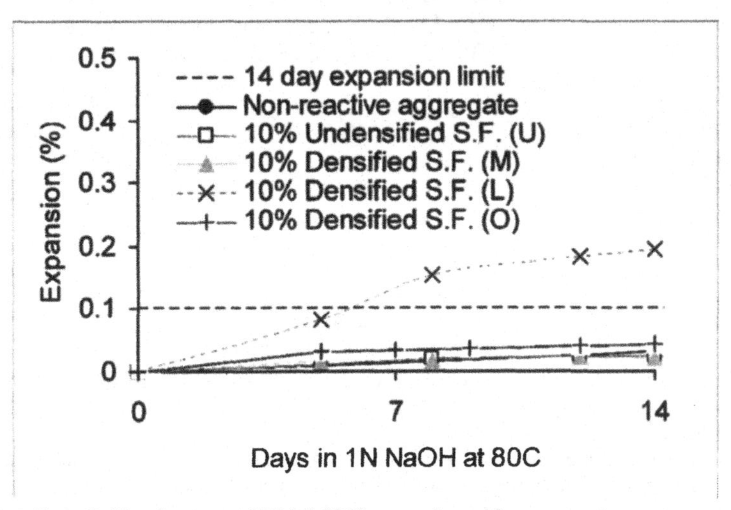

Figure 1: Effect of silica fume on ASTM C 1260 expansion with non-reactive aggregate

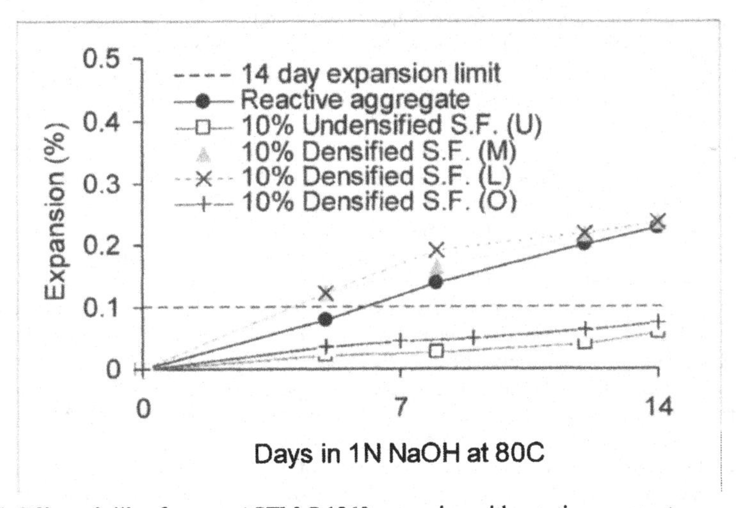

Figure 2: Effect of silica fume on ASTM C 1260 expansion with reactive aggregate

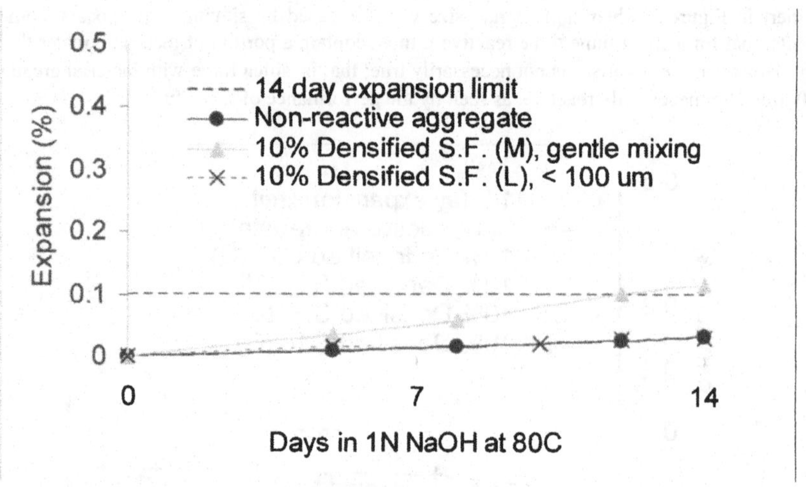

Figure 3: Effect of silica fume size on ASTM C 1260 expansion with non-reactive aggregate

CONCLUSION

Accelerated testing for ASR expansion with silica fume showed a variety of effects. Whereas undensified silica fume was inert with a non-reactive aggregate and reduced expansion of reactive aggregates, densified silica fume exhibited a variety of responses depending on its source; some increased expansion, some were inert, and some decreased expansion. The inconsistencies in the literature as to the behavior of silica fume agglomerates may therefore be explained by differences in materials tested across studies. It was observed that silica fume agglomerates greater than 100 µm are not necessarily reactive, but agglomerates less than 100 µm were always non-reactive for the materials tested.

REFERENCES

1. D. Bonen and S. Diamond, "Occurrence of Large Silica Fume-Derived Particles in Hydrated Cement Paste," *Cem. Concr. Res.*, **22**, 1059-1066 (1992).
2. G. Gundmundsson and H. Olafsson, "Silica fume in concrete – 16 years of experience in Iceland," in *Alkali-Aggregate Reaction in Concrete, Proceedings of the 10th International Conference,* Melbourne, 1996, pp. 469-562.
3. R.D. Hooton, R.F. Bleszynski, and A. Boddy, "Issues related to silica fume dispersion in concrete," in *Materials Science of Concrete – The Sidney Diamond Symposium*, American Ceramic Society, Westerville, Ohio, 1998, pp. 435-446.
4. P.R. Rangaraju and J. Olek, "Evaluation of the potential of densified silica fume to cause alkali-silica reaction in cementitious matrices using a modified ASTM C 1260 test procedure," *Cem. Concr. Agg.*, **22**, 150-159 (2000).
5. C. Perry and J.E. Gillott, "The feasibility of using silica fume to control concrete expansion due to alkali-aggregate reactions," *Durability of Building Materials*, **3**, 133-146 (1985).

6. A.M. Boddy, R.D. Hooton, and M.D.A. Thomas, "The effect of product form of silica fume on its ability to control alkali-silica reaction," *Cem. Concr. Res.*, **30**, 1139-1150 (2000).
7. H. Wang and J.E. Gillott, "Competitive nature of alkali-silica fume and alkali-aggregate (silica) reaction," *Mag. Concr. Res.*, **44**, 235-239 (1992).
8. M.C.G. Juenger and C.P. Ostertag, "Alkali–silica reactivity of large silica fume-derived particles," *Cem. Concr. Res.*, **34**, 1389-1402 (2004).
9. A.J. Maas, "Effects of Low-Level Silica Fume Replacement in ASTM C 1260 and ASTM C 227 Mortar-Bar Testing," Master's Thesis, The University of Texas at Austin, 2004.
10. ASTM C 1260-05, "Standard Test Method for Potential Alkali Reactivity of Aggregates (Mortar-Bar Method)," *Annual Book of ASTM Standards,* American Society for Testing and Materials, West Conshohocken, Pennsylvania, 2005.

MITIGATION OF ALKALI SILICA REACTION: A MECHANICAL APPROACH

Claudia P. Ostertag
Civil & Environmental Engineering Department
University of California
Berkeley, CA 94720, USA

ABSTRACT

The mechanical approach concentrates on modifying the mechanical environment in which the alkali silica reaction takes place. Matrices with different cracking and confinement characteristics are utilized to isolate the important parameters that influence gel initiation and alkali silica reaction rate. The paper is divided into two parts. Part I investigates the effect of crack control and confinement on alkali silica gel formation, reaction rate and gel composition using a matrix which exhibits crack growth resistance behavior. The crack growth resistance behavior confines the reactive aggregate and delays gel formation. In addition, the reaction products are being prevented from leaving the reaction site due to less cracking and small crack opening displacements which reduces the alkali silica reaction rate and modifies the gel composition. In Part II the matrices are designed to provide little confinement and allow the gel to expand almost freely. The matrices in Part II, however, differ in their ability to allow the gel to escape the reaction site. The results reveal that confinement causes a delay in gel formation whereas the lack of escape of gel products away from the reaction site reduces the alkali silica reaction rate.

INTRODUCTION

Alkali silica gel formation is caused by the chemical reactions between alkalis in Portland cement and the amorphous silica present in certain types of aggregates. Hydroxyl ions attack the silicon oxygen bond, creating an alkali-silica gel that is capable of imbibing water, resulting in volumetric expansion. This volumetric expansion causes cracking in cement-based materials if the expansion pressure exceeds the tensile capacity of the matrix.

The traditional approach to prevent alkali silica reaction (ASR) has focused on either preventing the ASR or reducing the expansion of the gel by modifying the chemical environment in which the alkali silica reaction takes place. Examples are: i) avoiding reactive aggregates, ii) using low alkali cement, or iii) either adding mineral admixtures[1-4] or lithium salts[5,6]. Avoiding reactive aggregates and limiting the alkali content in cement seem to be the easiest way to prevent damage in concrete due to ASR. However, quarries with adequate aggregates are being depleted and the use of non-ideal aggregates may be necessary in the future. Influencing the reaction chemistry by adding mineral admixtures to concrete had mixed results in terms of its effectiveness in reducing the expansion and hence damage associated with ASR. The mixed results are because of the wide variety of reactive aggregate types and sizes and cement types, and because of variations in chemical compositions[7] and agglomerations[8,9] of mineral admixtures. Therefore, it is of interest to investigate alternative forms of mitigating ASR.

This paper discusses a mechanical approach to reduce alkali silica reaction. The mechanical approach concentrates on designing matrices to resist gel formation and/or reduce alkali silica reaction rates. However, in order to mitigate ASR through the mechanical approach we need to know what governs gel initiation and alkali silica reaction rate in order to design matrices more effectively. Is initial confinement of the reactive aggregate essential in order to mitigate alkali silica reaction? Can we mitigate ASR by simply preventing the gel from leaving the reaction site without initial confinement? The paper provides answers to these questions by utilizing matrices with different cracking and confinement characteristics. Glass rods of constant diameter are being used as reactive aggregates which allow the gel formation and reaction rims to be compared among matrices with different confinement and cracking characteristics. The paper is divided into two parts. In part I the effect of crack control and confinement on alkali silica gel formation, reaction rate, and gel composition is investigated using a matrix which exhibits crack growth resistance behavior. In Part II the matrices are designed to provide little confinement on the gel and hence allow the gel to expand almost freely. However, the matrices in Part II differ in their ability to allow the gel to escape the reaction site. Results on the effect of the mechanical approach on alkali silica gel formation, reaction rate and gel composition will be discussed.

EXPERIMENTAL PROCEDURE

AS gel formation and reaction rates were studied using mortar specimens with various cracking and confinement characteristics. Since gel formation and rim thickness depends on the size and shape of the reactive aggregates, glass rods of constant diameter were chosen for these experiments in order to compare the rim formations and thicknesses between different mortar specimens. All specimens contain a glass rod of 5 mm in diameter as reactive aggregate embedded in their center. The mortar was placed in two layers, each with approx. 12mm in thickness. Prisms of 2.5 x 2.5 x 7.35cm were produced. Mortar specimens with the reactive glass rods and graded non-reactive aggregates were cast, cured in 80°C water bath for 1 day and immersed in a 1 N NaOH solution stored at 80°C following the ASTM C1260 procedure[10]. Exposing the specimens to a 1 N NaOH solution at 80°C accelerates the alkali silica reaction. The mortar matrices exhibit different cracking and confinement characteristics. For Part I the mortar matrix is reinforced with 0 and 7 vol% of steel microfibers (SMF), respectively. The SMFs have a rectangular cross-section of 20x100μm and a length between 3-5mm. Microfibers were used for the following reasons: i) due to their small size they can be in close vicinity to the reaction site and are right at the source where cracks will initiate; ii) they are able to bridge microcracks at onset before they become macrocracks, and iii) they exhibit a steep crack growth resistance behavior[11,12]. The aggregates in the mortar matrix were non-reactive limestone aggregates, graded according to ASTM C-1260. For Part II two different set of specimens were fabricated. Matrix of set I contains very reactive aggregates (i.e. the non-reactive limestone aggregates in matrix of Part I were replaced by the reactive aggregates). It was anticipated that the reactive aggregates will react with the alkalis before the glass rod does and hence cause cracking of the matrix prior to gel formation and cracking around the glass rods. These "pre-existing" matrix cracks will provide less confinement on the alkali silica gel that forms around the glass rod. In set 2, 5 mm long sections along the length of the glass rod contain tubes made out of polyvinylchloride (PVC). The PVC softens when exposed to the elevated temperatures associated with the ASTM C-1260 procedure, allowing the gel to expand freely. The specimen configurations for Part I and Part II are schematically shown in Fig. 1.

The prisms were used for microstructure analysis to investigate the onset of AS gel formation and its progression with increasing exposure time to NaOH solution. After each exposure time, the specimens were cut in their center, dried in desiccators and the cut surface embedded in epoxy. The epoxy was poured in vacuum to avoid possible air bubbles. The epoxy is cured at room temperature for 3 days instead of quick oven drying to minimize the thermal mismatch between epoxy and cement paste. The epoxy coated specimens were then polished with aluminum-carbide grit of size 600, and 15, 9, 3 and 1 μm diamond paste. Throughout the polishing process, kerosene was used as the fluid medium. The composition of the alkali-silica complex was analyzed by Inductive Coupled Plasma Spectroscopy (ICP).

RESULTS AND DISCUSSION

Part I

Effect of Crack Growth Resistance Behavior on Cracking Process and Expansion

Steel microfibers in cement mortars exhibit crack growth resistance behavior due to strengthening and toughening mechanisms associated with crack fiber interactions[11,12]. The difference in crack growth behavior between an unreinforced and SMF mortars is illustrated schematically in Fig. 2. At exposure time t_1, a crack initiates at the reactive aggregate/matrix interface in the unreinforced mortar specimens (Fig. 2a). The small resistance to crack extension in the unreinforced mortar is shown by the large increase in crack length and crack opening displacement with increasing exposure time, t, to the NaOH solution. The lack of crack growth resistance behavior in the unreinforced specimens reduces the driving force for crack extension, and hence smaller tensile stresses are sufficient to increase the crack length and the accompanying crack mouth opening displacement. Consequently, the effectiveness in confining the expansion of the reactive aggregate decreases due to decreasing compressive stresses acting on the expanding gel with increasing exposure time. The crack in the SMF mortars initiates at a higher expansion stress (i.e. at t_2 and not at t_1) and extends far less with increasing exposure time due to crack fiber interactions such as crack pinning, crack deflection and crack wake bridging (Fig. 2b). These energy absorbing mechanisms increase with increasing crack length due to the formation of a bridging zone behind the crack tip. Consequently, crack extension requires the tensile stresses to increase with increasing exposure time, contrary to the unreinforced specimen. The crack growth resistance behavior in SMF mortars, which requires higher tensile stresses for crack extension, increases the effectiveness of confining the expansion of the reactive aggregate with increasing exposure time to NaOH solution. This increase in confinement causes an increase in compressive stresses acting on the reactive aggregate.

Effect of Crack Control on Gel Formation and Reaction Rates.

Glass rods embedded in the SMF mortar matrices react far less compared to glass rods embedded in unreinforced mortar matrices for same exposure times to the NaOH solution. Fig.3 a and b are backscattered images of the remaining cross-sections of the glass rods for plain and SMF mortar specimens, respectively, exposed to NaOH solution for 35 days. The alkali silica reaction products form at the outer surface of the rod and the AS reaction continues towards the center of the glass rod. The dark regions seen between the remaining glass rod and the matrix in Fig 3a and b, respectively, will be referred to as the alkali silica reaction rims. These regions were originally filled with alkali silica gel, however, when the specimens were cut for processing, white precipitates and a watery liquid, presumably an alkali-silica complex, were lost resulting in a cavity between the remaining glass rod and the mortar. The cavity when filled with epoxy used for polishing the specimen surfaces shows up dark in the backscattered SEM images. The steel microfibers (show up white in the backscattered image) are evenly distributed and in close proximity to the reactive glass rod. Radial cracks that formed due to the expansive alkali silica reaction product are visible in Fig. 3a but are difficult to see in Fig. 3b at this low magnification due to their very small crack opening displacements.

The thickness of the reaction rims is plotted versus exposure time to NaOH solution in Fig. 4 for both the control and SMF specimens. In the control specimen the reaction rim forms early and increases in thickness with increasing exposure time. In the SMF specimens, no reaction rim has formed up to 13 days exposure to NaOH solution. Furthermore, the reaction rate is reduced considerably in SMF reinforced specimens compared to the control specimens. Permeability tests were performed to investigate if the delay in gel formation and reduced reaction rate is associated with a lower permeability due to the steel microfibers. The permeability tests were performed using rubidium hydroxide and the content of rubidium was analyzed using an electron microprobe. The results are plotted in Fig. 5. Despite its higher initial permeability the SMF reinforced specimens exhibit a delay in gel formation and lower reaction rates compared to the control specimens. Hence other factors seem to control the delay in gel formation and reduced reaction rates in SMF specimens.

Effect of Crack Control on AS Gel Composition

Table 1 presents results on the rim thickness and crack width as a function of exposure time to NaOH solution for the control and SMF specimens, respectively. Both the rim thickness and the crack width increase with increasing exposure time to NaOH solution. With increasing crack width the restraint on the glass rod is loosened. However, more importantly, more gel may be able to leave the reaction site and hence more gel can be produced, which may explain the higher reaction rate of the control specimens. In SMF specimens on the other hand, the reduced cracking and small crack opening displacements limits the migration of the ASR gel away from the reaction site into the surrounding mortar matrix. If the reactants cannot leave then the ion concentration has to increase. An increase in ion concentration would suppress the dissolution of the glass rod, thereby reducing the alkali silica reaction rate. To proof this concept, the ASgel liquid surrounding the glass rods (dark regions in fig. 1a and b) was collected in both the control and SMF specimens and its chemical composition analyzed by inductive coupled plasma (ICP) spectroscopy. Indeed both the Na and Si ion concentration in the SMF reinforced specimens is 33% and 45% higher than in the unreinforced specimens. The higher concentration can be attributed to the lack of escape of ASgel due to reduced crack formations and small crack opening displacements. The higher ion concentration suppresses further dissolution of the glass rod which explains the lower reactivity of the glass rod embedded in the SMF mortar specimens.

The experimental results of Part I reveal a delay in gel formation, a reduction in alkali silica reaction rate, and an increase in the ion concentration of the gel composition when cracking and crack opening displacements are minimized and the gel is being prevented from migrating away from the reaction site into the mortar matrix. However, in order to design matrices to mitigate alkali-silica reactions effectively, we need to know which of the two processes is dominating the mitigation, is it the confinement (i.e. compressive stresses acting on the reaction products) or is it the lack of escape of gel products away from the reaction site. Furthermore, is the initial confinement essential in order to mitigate alkali silica reaction or can ASR be reduced by simply preventing the gel from leaving the reaction site without initial confinement? Part II of the paper is designed to answer these questions.

Table 1: Average rim thickness and crack width as function of exposure time to 1 N NaOH solution for a) matrix with steel microfibers and b) unreinforced matrix.

a)

Days	Crack width (Sum), μm	Avg. Rim Thickness, μm	Stand. Dev., Rim, μm
6	41.9	56.5	99.3
13	103.9	41.9	36.2
20	198.5	142.1	96.0
27	176.3	156.9	84.9
37	287.1	178.5	35.0
41	195.2	249.3	139.9

b)

Days	Crack width (Sum), μm	Avg. Rim Thickness, μm	Stand. Dev., Rim, μm
6	NA	0	1.9
13	NA	0	9.1
20	NA	51.4	18.4
27	NA	16.2	30.5
37	39.2	46.6	9.7
41	52.5	74.7	27.9

Part II

Effect of Low Confinement on Gel Formation and Reaction Rate

Set 1a: In order to study the effect of low confinement on gel formation, the mortar matrix of set I needs to undergo considerable cracking before alkali silica reaction occurs around the glass rod. Microstructure analysis confirms the extensive cracking of the matrix before the glass rod shows any alkali silica reaction. Hence the effect of low confinement on gel formation around the glass rods could be investigated.

The rim thickness versus exposure time to NaOH solution is shown in Fig. 6 for set I and the SMF mortar specimens, respectively. The alkali silica gel formation initiates far earlier in set I compared to Part I with steel microfibers. No gel formation could be observed up to 13 days exposure to NaOH solution in glass rods surrounded by the SMF matrix whereas set I exhibits a reaction rim of 72 μm already after 6 days exposure to 1 N NaOH solution. This may be explained by the difference in confinement imposed on the AS gel. The pre-existing cracks in the matrix of set I prevent confinement of the gel. On the other hand, the gel in Part I remains confined up to longer exposure times due to the lack of cracking and crack opening displacements associated with the microfibers. Furthermore, the reaction rate is far higher in set 1 compared to the SMF specimens. The large opening of the matrix cracks allows the gel to leave the reaction site which may be responsible for the enhanced reaction rate.

Set 2: Sections along the length of the glass rod contain PVC tubes. When the specimens are immersed in 1 N NaOH solution held at 80°C, the PVC softens and allows the gel to expand freely. Again, similar to set 1 early gel formation is observed due to lack of confinement. Backscattered images taken after 7 days of exposure to NaOH solution reveal alkali silica reaction around the glass rod that contained PVC (Fig. 7a) but no AS gel could be observed along the periphery of the glass rod without PVC (Fig. 7b) . The low confinement due to the

PVC tube cause early gel formation similar to set 1. The rim thickness without PVC increases steadily with increasing exposure time as shown in Fig. 8. For sections with PVC the rim thickness increases at a higher rate compared to without PVC up to 20 days of exposure time. However, beyond the 20 days, the reaction rate slows down and the rate remains lower compared to sections along the glass rod without PVC. The reduced reaction rate at higher exposure times may be associated with the lack of escape of the reaction products. The PCV rod did not fracture and hence the gel is not able to escape.

Matrix Design to Mitigate ASR

The most effective way to mitigate ASR is accomplished in Part I with a matrix that exhibits crack growth resistance behavior. Not only is the AS gel formation delayed but also the AS reaction rate is considerably reduced. Hence matrices that delay crack initiation and minimize crack width will be most effective in mitigating ASR. Set 1 of Part II provides the least effective way to mitigate ASR. Set 1 is similar to many concrete structures which exhibit cracks due to plastic and/or drying shrinkage prior to ASR. The pre-existing cracks loosen the confinement on the alkali silica gel which allows the gel to form earlier compared to a confined case where the matrix exhibits crack growth resistance behavior. Furthermore, the alkali silica reaction rate is considerably enhanced due the ability of the gel to escape the reaction site due to the pre-existing cracks. Set 2 provides no confinement which allows the gel to form earlier similar to set 1. However, set 2 differs from set 1 in preventing the gel from escaping the reaction site. Despite lack of confinement, the alkali silica reaction rate and hence reaction products can be considerably reduced if they are prevented from leaving the reaction site. These results point out the importance of controlling cracking and crack opening displacements in order to mitigate alkali silica reaction.

CONCLUSION

A mechanical approach to mitigate alkali silica reaction was investigated. The mechanical approach concentrates on modifying the mechanical environment in which the alkali silica reaction takes place. Matrices with different cracking and confinement characteristics were utilized. The paper is divided into two parts.

Part I investigates the effect of crack control and confinement on alkali silica gel formation, reaction rate and gel composition. The experimental results reveal a delay in gel formation, a reduction in alkali silica reaction rate, and an increase in the ion concentration of the gel composition when cracking and crack opening displacements are minimized and the gel is being prevented from migrating away from the reaction site into the mortar matrix.

In Part II the matrices of set 1 and set 2 are designed to provide little confinement on the gel and hence allow the gel to expand almost freely. However, the matrices differ in their ability to allow the gel to escape the reaction site. The gel formation initiates at shorter exposure times to NaOH solution compared to Part I and is accelerated due to the low confinement on the AS gel. Furthermore, if reaction products are able to leave the reaction site as for set 1, far higher reaction rates are observed compared to Part I. On the other hand, if the reaction products can not leave as in set 2, the reaction rate decreases even for the case of low confinement. This emphasizes the importance of minimizing crack width in slowing down the alkali silica reaction rate.

In summary the initiation of gel formation is determined by the confinement. A high confinement delays gel formation whereas a low confinement accelerates gel formation. The reaction rate is determined by the ability of the gel to escape from the reaction site. A high reaction rate is observed when the gel is able to leave the reaction site through cracks of sufficient crack opening displacements. A low reaction rate is observed if cracking is prevented and/or the cracks have crack opening displacements small enough to prevent the escape of the gel from the reaction site. Hence matrices that minimize crack formation and reduce crack width will be very effective in mitigating alkali silica reaction.

ACKNOWLEDGEMENTS

The author would like to thank Hae-Young Song and Kevin Cheng for the sample preparations. Part of the funding was provided by the National Science Foundation Grant No. CMS-962480

REFERENCES

[1]W.Aquino, D.A.Lange, J.Olek, "The influence of metakaolin and silica fume on the chemistry of alkali-silica reaction products", *Cement and Concrete Composites* **23,** 485-493 (2001).

[2]S. Sprung, M. Adadian, "The effect of admixtures on alkali-aggregate reaction in concrete", Proc. Symp. Effect of alkalis on the properties of concrete, London, Sep. 1976. Cement and concrete association, Wexham Springs, Slough, pp.125-137.

[3]R.E. Oberholster and D.M. Roy,"The effectiveness of mineral admixtures in reducing expansion due to the alkali-aggregate reaction with Malmesbury group aggregates", *Proc. 5th Int.Conf.Alkali-aggregate reaction in concrete*, Cape Town, National Building Research Institute, Pretoria, 1981, paper S252/31.

[4]M.C.G. Juenger and C.P. Ostertag, "Alkali-silica reactivity of large silica fume-derived particles", *Cem.Conr.Res.* **34**, 1389-1402 (2004)

[5]E.J. MacCoy, A.G. Caldwell, "New approach to inhibiting alkali-aggregate expansion", *J. Am. Concr. Inst.* **22,** 693-706 (1951)

[6]S. Diamond, S. Ong, "The mechanisms of lithium effects on ASR", *Proceedings of the 9th International Conference on Alkali Aggregate Reaction*, Concrete Society of U.K., London, 269-278 (1992)

[7]D.W. Hobbs, "Deleterious expansion of concrete due to alkali-silica reaction: influence of PFA and slag," *Mag.Concr.Res.* **36,** 191-205 (1986).

[8]S. Diamond, "Alkali silica reaction – some paradoxes", *Cem Concr Comp* **19**, 391-401 (1997).

[9]D.A. St. John, S.A. Freitag, "Fifty years of investigation and control of AAR in New Zealand, in Alkali-Aggregate Reaction in Concrete", *Proceedings of the 10th International Conference,* Melbourne, 150-157 (1996).

[10]ASTM C 1260-94 (1999), Standard test method for potential alkali reactivity of aggregates (mortar-bar method) in Annual book of ASTM Standards v. 04.02

[11]C.P. Ostertag and CK. Yi, "Quasi-brittle behavior of cementitious matrix composites, " *Mat. Sci. and Eng.* **A 278**, 88-95 (2000)

[12]CK. Yi and C.P. Ostertag, "Strengthening and Toughening Mechanisms in microfiber reinforced cementitious composites", *J. Mat. Sci.* **36,** 1513-1522 (2001).

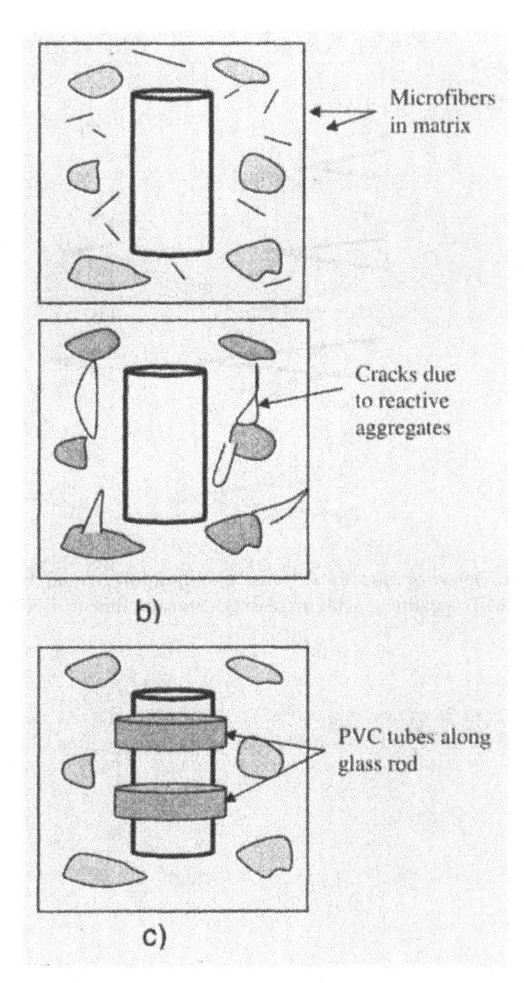

Fig.1: Schematic of specimens of Part I and Part II with reactive glass rod embedded in center; a) Part I: matrix surrounding glass rod contains microfibers and non-reactive limestone aggregates; b) Part II set 1: very reactive aggregates cause matrix to crack (after specimens are immersed in 1 N NaOH solution) prior to ASR around glass rod. c) Part II set 2: glass rod with PVC tubes along its length; matrix contains non-reactive limestone aggregates

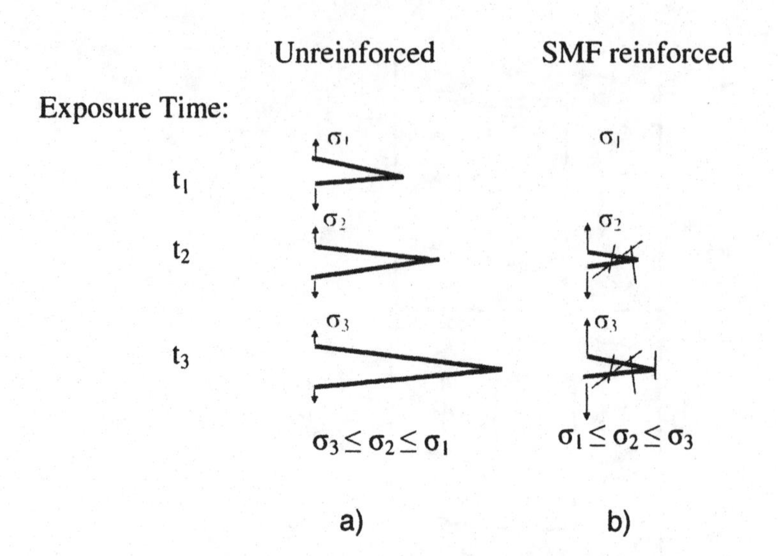

Fig.2: Crack growth behavior due to ASR in a) unreinforced and b) steel microfiber reinforced (SMF) specimens with increasing exposure time to 1 N NaOH solution

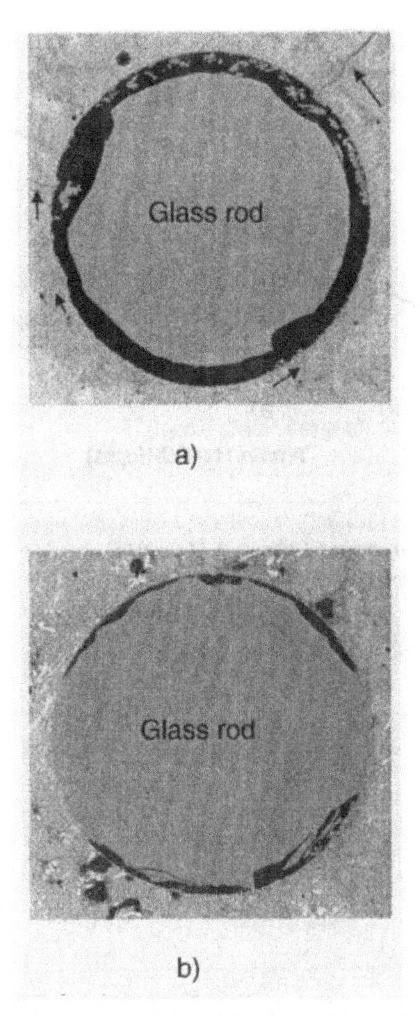

Fig.3: Backscattered (SEM) images of the remaining cross section of the glass rod after exposure to 1 N NaOH (aq) for 35 days (actual area size of 6 x 6mm); a) unreinforced specimens, arrows point to cracks that formed to the ASR; b) steel microfiber reinforced specimen;

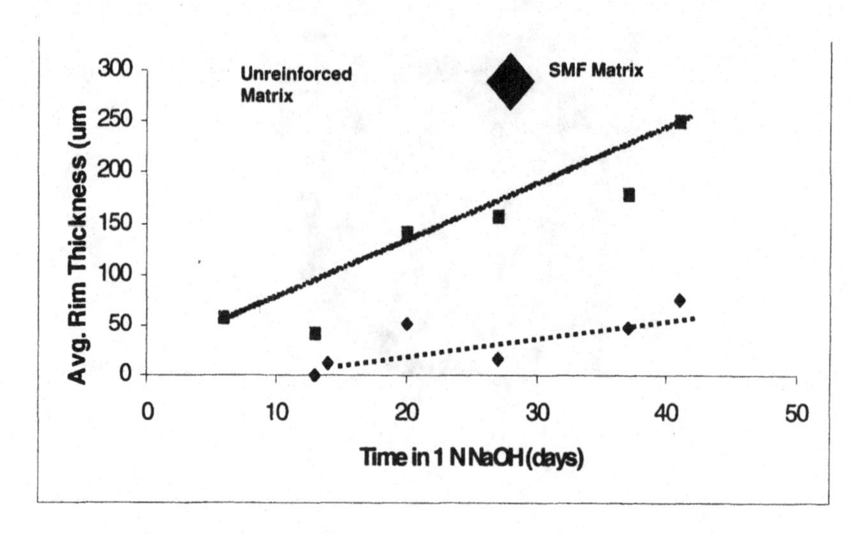

Fig.4: Average Rim thickness versus time in 1 N NaOH solution for unreinforced matrix and steel microfiber reinforced matrix.

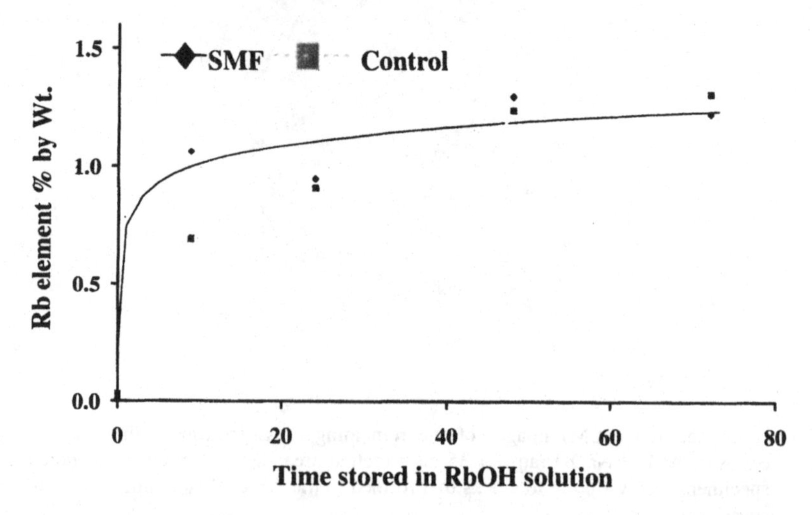

Fig.5: Permeability results for steel microfiber reinforced (SMF) and unreinforced (control) specimens.

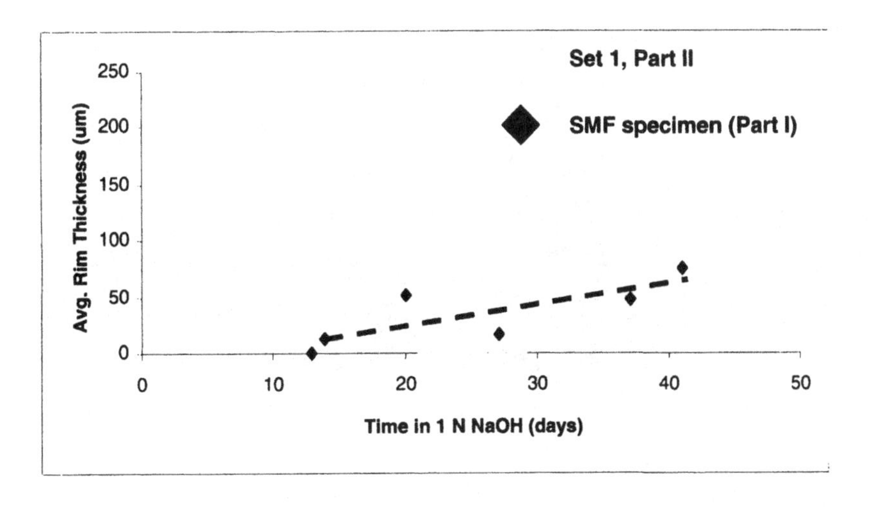

Fig.6: Average Rim thickness versus exposure time in 1 N NaOH solution for steel microfiber reinforced matrix (Part I) and set I (Part II).

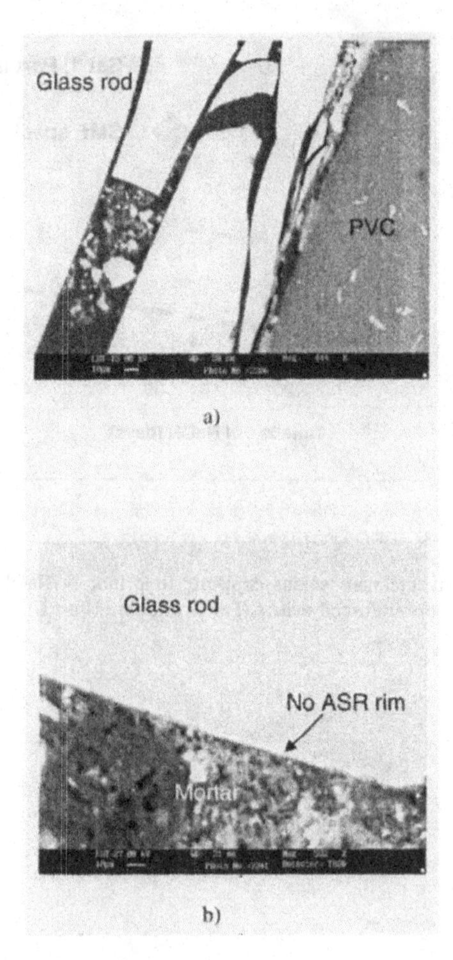

Fig.7: Backscattered SEM images after 7 days of exposure to NaOH solution; a) with PVC tube, b) without PVC tube.

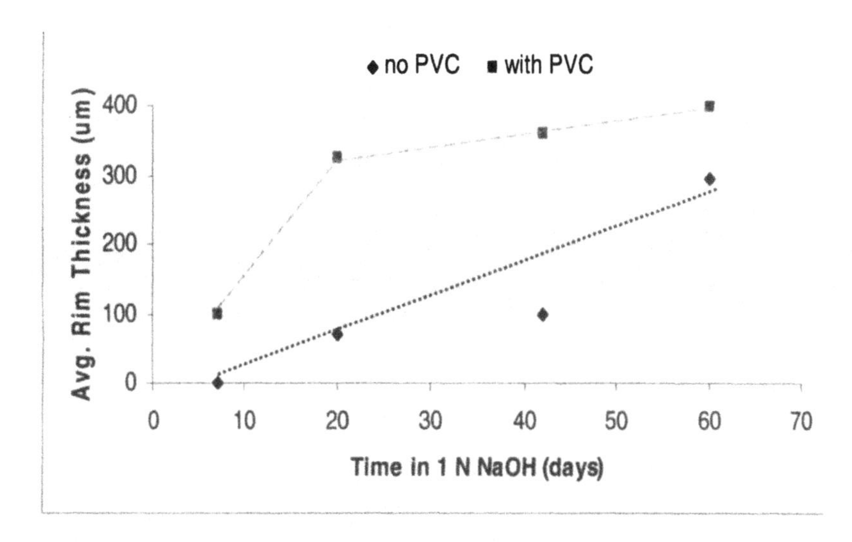

Fig. 8: Average Rim thickness versus exposure time in 1 N NaOH solution for set 2, Part II, along glass rod with and without PVC tube.

EXPERIMENTAL STUDIES AND PARALLEL COMPUTING ON COUPLED MOISTURE AND CHLORIDE DIFFUSION IN CONCRETE

[1]Yunping Xi, [2]Ayman Ababneh, [1]Lydia Abarr, [1]Suwito, and [1]Xiao-Chuan Cai
[1]University of Colorado, Boulder, CO 80309
[2]Clarkson University, Potsdam, NY 13699

INTRODUCTION

The interaction between moisture diffusion and chloride penetration in concrete affects the durability of reinforced concrete structures. The corrosion of the reinforcement in concrete takes place when the chloride content of concrete near steel bar has reached a threshold value and the moisture content in the concrete is sufficiently high. Therefore, moisture and chloride ions are two necessary conditions for the onset of corrosion of rebars in concrete. The diffusions of chloride and moisture in concrete can be studied for two different situations: fully saturated and partially saturated. In the first situation, the concrete is fully saturated, and the dominant mechanism for both chloride diffusion and moisture diffusion is the concentration gradient of chloride. In another word, the chloride concentration gradient drives not only the chloride penetration but also the moisture movement in the concrete. In the second situation, the concrete is partially saturated, and the moisture concentration gradient (in addition to the chloride concentration gradient) results in the moisture penetration as well as the chloride diffusion. In this case, both concentration gradients are driving forces.

Systematic experimental studies have been performed at University of Colorado at Boulder to study the coupling effects between the moisture diffusion and chloride penetration in concrete. Ababneh and Xi (2002) conducted an experimental study on the first situation (saturated concrete with chloride gradient) in which the coupling effect of chloride penetration on moisture diffusion was studied experimentally. The results showed that the coupling effect is significant, and it can be characterized by a coupling parameter, $D_{H\text{-}Cl}$. The experimental data were used to determine $D_{H\text{-}Cl}$ and the results showed that $D_{H\text{-}Cl}$ is not a constant but depends on the chloride concentration. Most recently, Abarr (2005) conducted an experimental study for the second situation: to investigate the effect of moisture diffusion on chloride penetration in non-saturated concrete. The experimental results showed that the coupling effect is significant, and it can be characterized by another coupling parameter, $D_{Cl\text{-}H}$. The experimental data were used to determine $D_{Cl\text{-}H}$ and the results showed that, similar to $D_{H\text{-}Cl}$, $D_{Cl\text{-}H}$ is not a constant but depends on the chloride concentration.

When the two coupling parameters $D_{H\text{-}Cl}$ and $D_{Cl\text{-}H}$ with concentration dependency are included in the diffusion equations for moisture and chloride transport in concrete, the equations become coupled nonlinear partial differential equations, and numerical approaches such as finite element analysis must be employed. Both moisture diffusion and chloride diffusion in concrete take place only in a thin surface layer of the structure, resulting in steep gradients of the diffusing species. Consequently, very fine finite element mesh is needed for the diffusion analyses. Steep gradients of the diffusing species in turn cause high stress concentration in the structure, for example, high shrinkage stress is resulted from drying shrinkage in surface layer. As a result, finite element models for performance evaluation of diffusion-induced damage in existing structures are usually very large. Thousands and even millions of elements are needed depending on the size and geometry of the structure. The computational task becomes very challenging. One possible solution is to use parallel computing technique. Large structural models can be divided by regions or by structural members (e.g. beams and columns), and each region or member can be handled by one processor in the parallel computing system. Hence, overall computing time for a large-scale diffusion analysis can be reduced significantly. To this end, a parallel computing algorithm was developed for coupled processes of moisture diffusion and chloride penetration in concrete (Suwito et al. 2006), and it is now being used for diffusion analysis of large scale bridge structures in Colorado.

This paper provides a detailed overview on recent research results obtained at University of Colorado at Boulder. Some of the results were published elsewhere and some of them have not been published. Theoretical formulations related to the two coupled diffusion processes will be described first, followed by the experimental studies on the two coupling effects and the determination of the two coupling parameters. Then, the parallel computing algorithm will be presented together with a numerical example on a concrete slab subjected to simultaneous moisture diffusion and chloride penetration.

BASIC FORMULATION

Usually, the diffusion of moisture and the diffusion of chloride in concrete are described by conventional Fick's first law: the flux of moisture, J_H, is proportional to the moisture gradient or the gradient of pore relative humidity in concrete, H; and the flux of chloride, J_{Cl}, is proportional to the concentration gradient of free chloride C_f

$$J_H = -D_{H-H}\, grad(H) \tag{1}$$

$$J_{Cl} = -D_{Cl-Cl}\, grad(C_f) \tag{2}$$

In Eqs. (1) and (2), there are two material parameters. D_{H-H} is the humidity diffusion coefficient, and D_{Cl-Cl} the chloride diffusion coefficient. Based on mass conservation and Eqs. (1) and (2), two partial differential equations can be obtained (the so-called Fick's second law):

$$\frac{\partial w}{\partial t} = -div(J_H) = div\left[D_{H-H}\, grad(H)\right] \tag{3}$$

$$\frac{\partial C_t}{\partial t} = -div(J_{Cl}) = div\left[D_{Cl}\, grad(C_f)\right] \tag{4}$$

in which w is the water content in concrete, and C_t the total chloride content in concrete including the free and the bound chloride.

Now, we consider the chloride and moisture gradients as the two driving forces acting simultaneously in the concrete. The conventional Fick's first law may be modified to take into account the two driving forces. The flux of moisture in concrete is assumed to consist of the flux due to the gradient of moisture concentration as well as the flux due to the gradient of free chloride (Ababneh and Xi 2002); and similarly, the flux of chloride ions is composed of a flux due to chloride concentration gradient and the flux due to moisture concentration gradient (Abarr 2005)

$$J_H = -D_{H-H}\, grad(H) - D_{H-Cl}\, grad(C_f) \tag{5}$$

$$J_{Cl} = -D_{Cl-Cl}\, grad(C_f) - D_{Cl-H}\, grad(H) \tag{6}$$

In Eqs. (5) and (6), there are four material parameters. D_{H-H} and D_{Cl-Cl} have the same meanings as in Eqs. (1) and (2); D_{H-Cl} is the coupling parameter for the effect of chloride penetration on moisture diffusion, and D_{Cl-H} the coupling parameter for the effect of moisture diffusion on chloride penetration. Eqs. (5) and (6) may be considered as two modifications to the conventional Fick'law to included the two coupling effects between moisture and chloride diffusion. Some similar proposals were made previously to modify Fourier law and Fick's law when the coupling effects between moisture transfer and heat conduction in porous media were considered. For example, J_H is assumed to be the sum of the flux due to the gradient of moisture concentration H (Fick's law of diffusion) and the flux due to the gradient of temperature T (Soret flux); and the heat flux, J_T, is composed of a flux due to temperature gradient (Fourier law) and the flux due to moisture concentration gradient (Dufour flux).

Based on mass conservation and Eqs. (5) and (6), two coupled partial differential equations can be obtained:

$$\frac{\partial w}{\partial t} = \frac{\partial w}{\partial H}\frac{\partial H}{\partial t} = div\left[D_{H-H}\, grad(H) + D_{H-Cl}\, grad(C_f)\right] \tag{7}$$

$$\frac{\partial C_t}{\partial t} = \frac{\partial C_t}{\partial C_f}\frac{\partial C_f}{\partial t} = div\left[D_{Cl-Cl}\, grad\left(C_f\right) + D_{Cl-H}\, grad\left(H\right)\right] \tag{8}$$

There are two new material parameters in Eqs. (7) and (8), $\partial w/\partial H$ and $\partial C_t/\partial C_f$. $\partial w/\partial H$ is the moisture capacity and $\partial C_t/\partial C_f$ the chloride binding capacity. Comparing the coupled diffusion equations, Eqs (7) and (8) with un-coupled equations, Eqs. (3) and (4), one can clearly see that the solutions for the coupled equations are much more difficult to obtain.

The two diffusion coefficients and the two capacities in Eqs. (7) and (8) have been studied extensively (Tang and Nilsson 1992; Saetta et al. 1993; Andrade and Sanjuan 1994; Xi et al. 1994a; Xi et al. 1994b; Xi et al. 1995a; Xi et al. 1995b; Wee et al. 1997; Xi and Bazant 1999; Xi et al. 2000; Swaddiwudhinpong et al. 2000; Ababneh et al. 2003; Xi and Nakhi 2005), and they will not be described in this paper. The focus of this paper will be the two coupling parameters D_{H-Cl} and D_{Cl-H}. The experimental study on D_{H-Cl} will be presented first and then the experimental study on D_{Cl-H}. After the determination of the two coupling parameters, a finite element algorithm will be presented for solving Eqs. (7) and (8) by a parallel computing technique.

AN EXPERIMENTAL STUDY ON THE COUPLING PARAMETER D_{H-Cl}

The first purpose of this experimental study is to examine the significance of the effect of chloride diffusion on moisture movement. If the effect is not significant, the second terms on the right hand side of Eq. (5) and Eq. (7) can be omitted, and then the moisture diffusion equation, Eq. (7) would be un-coupled. On the other hand, if the effect is very significant, the second terms on the right hand side of Eq. (5) and Eq. (7) cannot be neglected. In this case, the coupling parameter D_{H-Cl} will be determined based on the present test data. Thus, the second purpose of this study is to develop a procedure that can be used to evaluate D_{H-Cl} based on the test data.

Fig. 1 - Experimental setup for the effect of chloride diffusion on moisture diffusion.

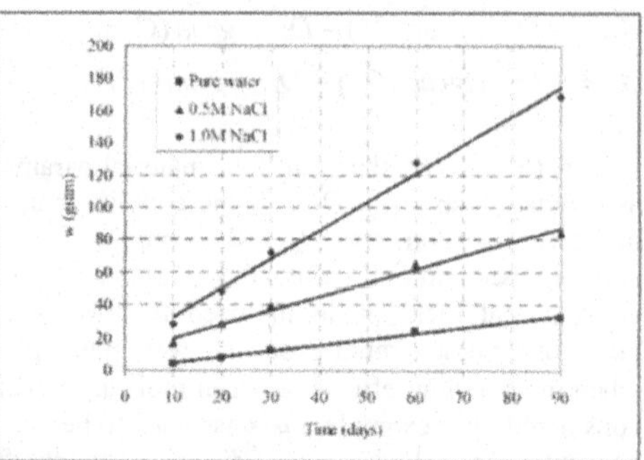

Fig. 2 - Exposure time versus the weight of accumulated solution in the center holes.

The concrete specimen used for this study was a cube (20x20x15 cm) with a central hole (10cm in diameter and 11.4 cm in depth). The samples were casted, demolded after 24 hours and stored in a moist room for 28 days. The experimental setup is shown in Fig. 1. The proportions of the concrete mixes used in the experiments were weight of water 213 *(Kg/m^3)*, cement 380 *(Kg/m^3)*, sand 702 *(Kg/m^3)*, and gravel 911 *(Kg/m^3)*. The water-cement ratio was 0.56. The average 28-day compressive strength was 34.1 *MPa*. Three sets of concrete specimens were immersed in three bathes containing three different solutions. Each set had three concrete specimens. All specimens were saturated by water with internal relative humidity of 100%. The external surface of the center hole was exposed to the atmosphere (40% relative humidity) during the test.

Basically, this is a seepage test. Bath 1 contained water (on the right in Fig. 1), and thus the moisture gradient was the only driving force for the moisture diffusion in concrete. Bath 2 and Bath 3 (in the middle and left in Fig. 1) contain sodium chloride (NaCl) solutions with concentrations of 0.5M and 1.0M, respectively. Two driving forces co-exist in the specimens in Bath 2 and Bath 3: moisture gradient and chloride concentration gradient. The accumulated solution in the center holes was collected after 10, 20, 30, 60 and 90 days of ponding. Except at the time of collection for the accumulated solution, the ponding containers were covered all the time to reduce the evaporation of the accumulated solution in the hole.

The amount of water accumulated in the center holes of the specimens was used as a measure for the effect of chloride penetration on moisture diffusion. If there is a significant difference in the amounts of water accumulated in the center holes of specimens in Bath 1 and Bath 2, the difference must be due to the chloride concentration gradient; and this means that the coupling effect is significant. Moreover, if there is a significant difference in the amounts of water accumulated in the center holes of specimens in Bath 2 and Bath 3, the difference must be due to the difference of the chloride concentrations (0.5M and 1.0M), this means that the coupling effect is concentration dependent.

Fig. 2 shows the averaged experimental results. The most important observation is that the appearance of chloride ion gradient accelerates the moisture diffusion greatly! The amount of accumulated water is about tripled from zero chloride concentration to 0.5M, and then doubled from 0.5M to 1.0M. It is an important experimental evidence that the effect of chloride penetration on moisture diffusion is very significant.

Next, the coupling parameter can be determined by using the experimental data and Eq. (9) (see Ababneh and Xi 2002 for detailed derivation):

$$D_{H-Cl} = \frac{L}{A * \Delta C_f}\left[\left(\frac{\partial w}{\partial t}\right)^{Total} - \left(\frac{\partial w}{\partial t}\right)^{Moisture}\right] \quad (9)$$

in which ΔC_f is the difference in the chloride concentration between the bath solution and the solution accumulated in the central hole; A is the internal surface area of the hole in the specimen; L is the average thickness of the concrete wall surrounding the centre hole; $(\partial w / \partial t)^{Total}$ is the rate of solution accumulation in the central hole due to both moisture diffusion and chloride penetration (Bath 2 or Bath 3), and $(\partial w / \partial t)^{Moisture}$ is the rate of solution accumulation in the central hole due only to moisture diffusion (Bath 1). Using Eq. (9) and the test data from Bath 1 (water) and Bath 2 (0.5M concentration), we obtain $D_{H-Cl} = 1.423 g/(cm.day)$. Using Eq. (9) and the test data from Bath 1 (water) and Bath 3 (1.0M concentration), we have $D_{H-Cl} = 2.622 g/(cm.day)$. This means that the coupling parameter is not a constant but depends on chloride concentration. Within the concentration range of the present study, we may use a linear equation

$$D_{H-Cl} = \frac{D^0_{H-Cl}}{C^0_f}\Delta C_f = 2.846\Delta C_f \quad (10)$$

in which D^0_{H-Cl}/C^0_f represents the ratio of the reference coupling parameter at the reference concentration. When 0.5M is taking as the reference concentration, 2.846 is the ratio. It is important to note that the present test data were obtained only from one concrete mix design with high water-cement ratio, 0.56. More experimental studies are needed to verify Eq. (10) and to determine D^0_{H-Cl}/C^0_f as a function of concrete mix design parameters.

AN EXPERIMENTAL STUDY ON THE COUPLING PARAMETER D_{Cl-H}

The first purpose of this study is to examine the significance of the effect of moisture diffusion on chloride penetration. If the effect is not significant, the second terms on the right hand side of Eq. (6) and Eq. (8) can be omitted, and then the chloride diffusion equation, Eq. (7) would be un-coupled. On the other hand, if the effect is very significant, the second terms on the right hand side of Eq. (6) and Eq. (8) cannot be neglected. In this case, the coupling parameter D_{Cl-H} will be determined based on the present test data. Thus, the second purpose of the study is to develop a procedure that can be used to evaluate D_{Cl-H} based on the test data.

The basic idea of the experimental study is shown in Fig. 3, which is a ponding test. In Fig. 3(a), the concrete specimen is initially partially saturated and exposed to a NaCl solution on the top surface. Since the concrete is not fully saturated, there are two driving forces in the specimen for the moisture and chloride diffusion. In Fig. 3(b), the concrete specimen is initially fully saturated and exposed to the same NaCl solution on the top surface and to water on the bottom surface. Since the concrete is fully saturated all the time, there is only one driving force in the specimen for the chloride diffusion (neglecting gravity force). After a certain ponding period, the chloride concentration profiles in the two concrete specimens are measured and compared. If the profile in specimen (a) is higher than that in specimen (b), the difference must be due to the moisture gradient in specimen (a). The same ponding test can be used for concrete specimens exposed to chloride solutions of different concentrations (3% and 5% were used in this study). The concentration profiles will be compared. If there is any difference in the profiles, it means that the coupling effect is concentration dependent.

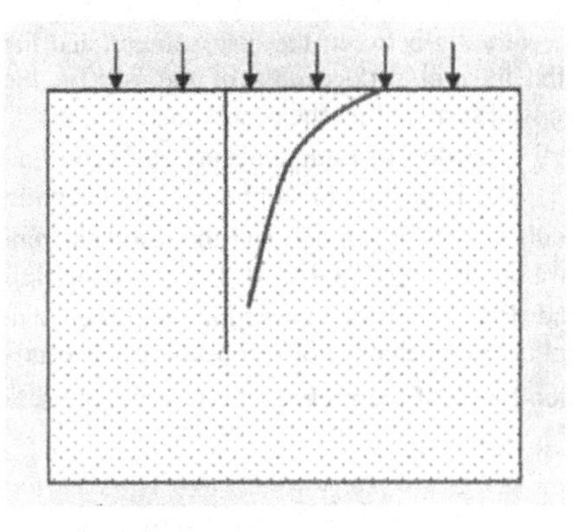

(a) Initially partially saturated specimen
the top surface is exposed to NaCl solution

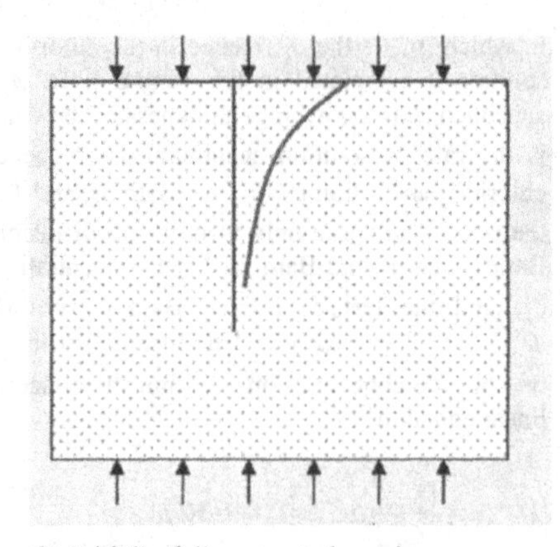

(b) Initially fully saturated specimen
the top surface is exposed to NaCl solution
and the bottom surface is exposed to water

Fig. 3 The experimental setup for the ponding tests

The samples were designed as four inch diameter cylinders with heights of two inches and three inches, respectively. The mix proportions of the concrete used in this study were the same as those listed above for the study on $D_{H\text{-}Cl}$. The partially saturated concrete samples were embedded with SHT75 Sensirion humidity and temperature sensors. These sensors were used to measure humidity profiles in the concrete, which provide necessary information for determining the coupling parameter. The sensors were wrapped in GorTex before the concrete was poured around them. This was to protect the sensors as well as allow a breathable interface so that the humidity readings would be accurate. Two sensors were placed in each sample, the first sensor was placed in the center, two inches from the solution surface, and the second sensor was placed one inch off center, one inch from the solution surface. The samples were cast, de-molded after 24 hours and stored in the moist room for 28 days. Then, some of the samples were kept under fully saturated state, and they were considered as initially fully saturated specimens. Other samples were kept in the lab until the internal relative humidity reduces to a constant, and these specimens were considered as initially partially saturated specimens.

The chloride penetration profiles were obtained after different ponding periods of 30 days, 60 days, and 90 days. Fig. 4(a) shows the profiles of 30 days ponding with 3% NaCl solution. One can clearly see that the chloride concentrations are higher in the initially partially saturated concrete than the initially fully saturated concrete, at the same location. This means that the moisture diffusion accelerates the chloride diffusion when the two diffusion processes occur in the same direction. Fig. 4(b) shows the profile of 30 days of ponding with 5% NaCl solution. One can see the same trend as in Fig. 4(a). These two figures confirm that there is a significant effect of moisture diffusion on chloride penetration.

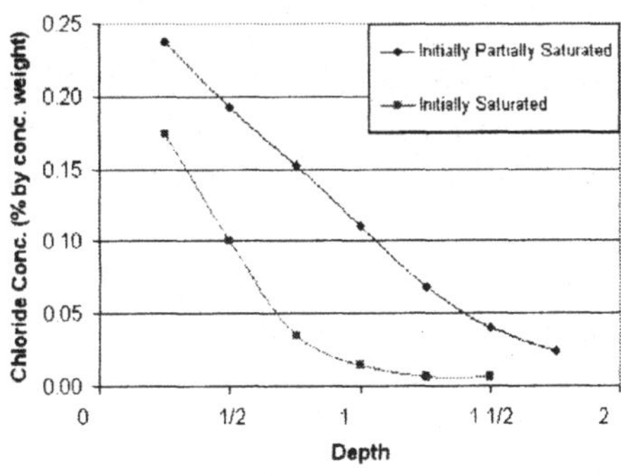

Fig. 4(a) Chloride profiles after 30 Days of exposure to 3% NaCl solution.

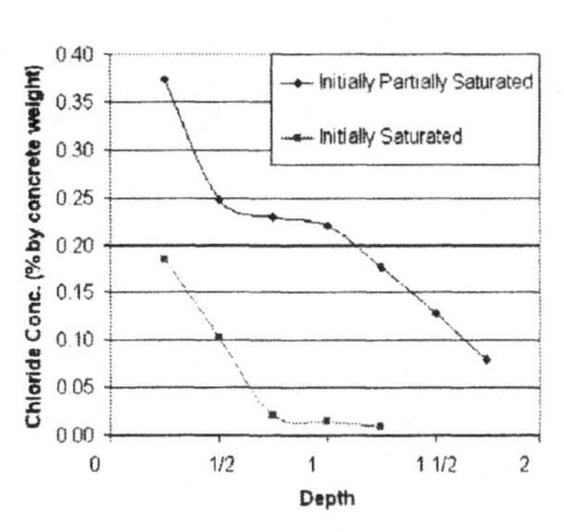

Fig. 4(b) Chloride profiles after 30 Days of exposure to 5% NaCl solution.

Furthermore, comparing Fig. 4(a) and Fig. 4(b), one can see that the difference of the two curves in Fig. 4(b) is larger than the difference in Fig. 4(a). This means that the coupling effect is due not only to the co-existence of the two driving forces but also to the concentration of chloride (Fig. 4(b) is from 5% concentration).

Next, we derived the formula to evaluate D_{Cl-H} based on the test data (detailed derivation can be seen in Abarr 2005):

$$D_{Cl-H} = \left(\frac{\partial C_t}{\partial t} - \frac{\partial C_{Cl}}{\partial t} \right) \frac{\Delta x}{A \Delta H} \tag{11}$$

where $(dC_t/dt)^{partial}$ is the change of chloride concentration at a fixed depth due to two driving forces (i.e. in initially partially saturated concrete); $(dC_t/dt)^{full}$ is the change of chloride concentration at a fixed depth due to one driving force (i.e. in initially fully saturated concrete), Δx is the distance between two adjacent depths; A is the area of sample surface; and ΔH is the internal relative humidity readings at two different locations at the same time. Fig. 5 shows the distribution of the internal relative humidity at 30 days. The coupling parameter, D_{Cl-H}, was calculated using the data for the first 30 days.

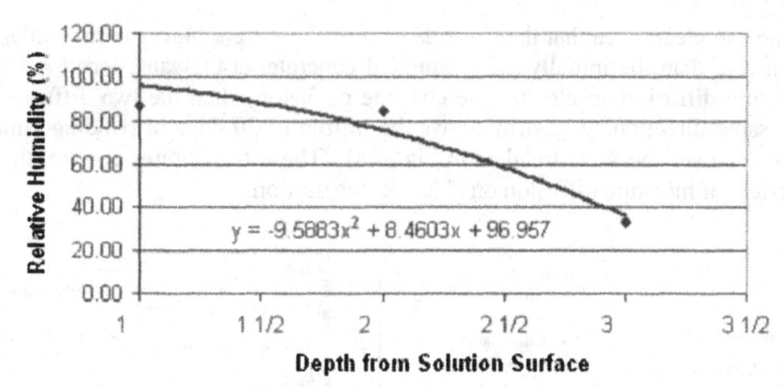

Fig. 5 Relative humidity vs. depth from the concrete exposed to 3% Solution

Using Eq. (11) and the test data of chloride profiles and humidity profiles, the coupling parameter D_{Cl-H} can be evaluated. Fig. 6 shows the results. As one can see D_{Cl-H} is concentration dependant, and it increases linearly with chloride concentration. A linear equation can be obtained from Fig. 6.

$$D_{Cl-H} = \frac{D^0_{Cl-H}}{C^0_f} C_f = \frac{2.8\times10^{-11}(g/cm.\sec)}{0.1(\%)} C_f = 2.8\times10^{-10} C_f (g/cm\cdot\sec) \tag{12}$$

From Eq. (12), we have $D_{Cl-H} = 0$ when $C_f = 0$. This is an important and necessary condition for the flux equation of chloride, Eq. (6). When there is no chloride in concrete, the chloride flux must be zero, $J_{Cl} = 0$, which means the right hand side of Eq. (6) must zero. When $C_f = 0$ everywhere, $grad(C_f) = 0$, hence the first term of the right hand side of Eq. (6) equals zero. For non-saturated concrete, $grad(H) \neq 0$, therefore, the second term of the right hand side of Eq. (6) equals zero only if $D_{Cl-H} = 0$ at $C_f = 0$. This is evidenced by the test data in Fig. 6 and Eq. (12).

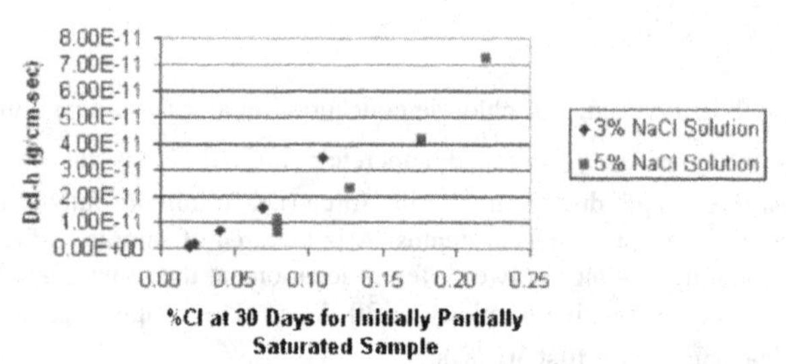

Fig. 6 D_{Cl-H} vs. chloride concentration for samples exposed to 5% solution

Again, it is important to note that the present test data were obtained only from one concrete mix design with high water-cement ratio, 0.56. More experimental studies are needed to verify Eq. (12) and to determine D^0_{Cl-H}/C^0_f.

PARALLEL COMPUTING TECHNIQUE FOR THE COUPLED DIFFUSION EQUATIONS

Eqs. (7) and (8) are the governing equations for the coupled problem of chloride ions and moisture diffusions, and they can be written in the following format in terms of free chloride concentration, C_f, and pore relative humidity, H

$$\frac{\partial C_t}{\partial C_f}\frac{\partial C_f}{\partial t}=\nabla\cdot\left[D_{Cl-Cl}\nabla C_f+D_{Cl-H}\nabla H\right] \tag{13}$$

$$\frac{\partial w}{\partial H}\frac{\partial H}{\partial t}=\nabla\cdot\left[D_{H-Cl}\nabla C_f+D_{H-H}\nabla H\right] \tag{14}$$

It should be noted that we did not use Eqs. (10) and (12) obtained in the experimental studies for the two coupling parameters, because the reference values in Eqs. (10) and (12) were obtained from our concrete specimens with high water-cement ratio, while the material models for the other four parameters, the two capacities and two diffusivities, in Eqs. (13) and (14) were developed not based this concrete mix design. As an approximation, we employed $D_{Cl-H} = \varepsilon D_{H-H}$ and $D_{H-Cl} = \delta D_{Cl-Cl}$ to take into account the coupling effects. ε = humidity gradient coefficient, which represents the coupling effect of moisture diffusion on chloride penetration. In some literatures on transport properties of concrete, D_{Cl-Cl} is called absolute chloride diffusivity, and $D_{Cl-H} = \varepsilon D_{H-H}$ is called relative chloride diffusivity. Similarly, δ = chloride gradient coefficient, which represents the coupling effect of chloride ions on moisture diffusion. D_{H-H} may be called absolute humidity diffusivity, and $D_{H-Cl} = \delta D_{Cl-Cl}$ may be called relative humidity diffusivity. ε and δ are two constants. As pointed out earlier, more experimental studies are definitely needed to improve Eqs. (10) and (12) so that they can be used in future theoretical modeling.

The general boundary conditions for Eqs. (13) and (14) are

$$C_f=C_0 \text{ on } \Gamma_1 \tag{15}$$

$$D_{Cl-Cl}\frac{\partial C_f}{\partial n}+J_{Cl}+D_{Cl-H}\frac{\partial H}{\partial n}+\alpha_{Cl}\left(C_f-C_{fa}\right)=0 \text{ on } \Gamma_2 \tag{16}$$

$$C_f=C_0 \text{ on } \Gamma_1 \tag{17}$$

$$D_{H-H}\frac{\partial H}{\partial n}+J_H+D_{H-Cl}\frac{\partial C_f}{\partial n}+\alpha_H\left(H-H_a\right)=0 \text{ on } \Gamma_4 \tag{18}$$

where α_{Cl} and α_H are the convective chloride and relative humidity coefficients; and C_{fa} and H_a are ambient chloride ions and relative humidity; respectively. Γ_1 and Γ_3 are the part of boundary with a constant chloride ions and relative humidity; and Γ_2 and Γ_4 are the part of boundary subjected to a specified chloride ions and relative humidity flux; respectively. Γ_1 and Γ_2 form the complete boundary surface for the chloride ions diffusion problem, and Γ_3 and Γ_4 for the moisture diffusion problem.

A parallel computing program was then developed for Eqs. (13) and (14) with boundary conditions of Eqs. (15) to (18) (Suwito et al. 2006). As discussed in the introduction, the purpose of developing the parallel computing program is to make a significant reduction in the executing time of the program. Ideally, if the executing time for solving Eqs. (13) and (14) is *T* on a one-processor system, then, with a *p*-processor system, the executing time would be simply *T/p*. But, in reality, the computational work cannot be evenly distributed within a number of processors without the need of any communication between the processors. Therefore, the ideal situation almost does not exist. The goal of an efficiently designed parallel program is to achieve the highest reduction in the executing time. The parallel implementation of the finite element program developed in this study is based on the distributed memory system, in which each processor has its own memory module. Such a distributed memory system is built by connecting each component with a high-speed communications network. Individual processor communicates to each other over the network. Inter-processor communications are managed by a message-passing technology, called Message Passing Interface (MPI). Developing parallel program using MPI directly is quiet complex. Thus, PETSc is used in this study, which is a higher level library than MPI and it provides a suite of data structures and routines for the scalable (parallel) solution of Eqs. (13) and (14).

In general, there are two strategies that can be used for parallel implementation of a finite element program, i.e.: (1) the system equations are formed by the usual sequential approach (without any partitioning of problem domain), but solved in parallel; and (2) the problem domain is divided into a number of sub-domains and then the system equations are formed concurrently for each sub-domain and solved also in parallel. The first approach, also called implicit domain decomposition approach, is generally suitable for static or steady state and linear problems; while the second approach, called the explicit domain decomposition approach, is more suitable for time dependent and nonlinear problems. Since Eqs. (13) and (14) are nonlinear and time dependent, the second approach is employed. There are two types of domain decomposition methods, overlapping and non-overlapping methods. The non-overlapping method, which is also known as sub-structuring method, is employed in this study.

The coupled diffusion equations were solved using the parallel computing program and the results of model simulation were compared with available test data. Some of the material parameters in the material models were adjusted in order to obtain the best prediction of the concentration profiles (Suwito et al. 2006). Then, a numerical example was used to show the computational results. It is a concrete slab of 15 cm by 30 cm, shown in Fig. 7. The concrete slab contains initially no chloride ions and it has 60% relative humidity (RH). The concrete slab is exposed to 3% NaCl and 100% RH on the top surface. All other boundaries are assumed to be sealed. This example is not to simulate the concrete under the service condition, but to simulate the commonly used long-term pond test. There are two driving forces for the diffusion of chloride and moisture in the concrete.

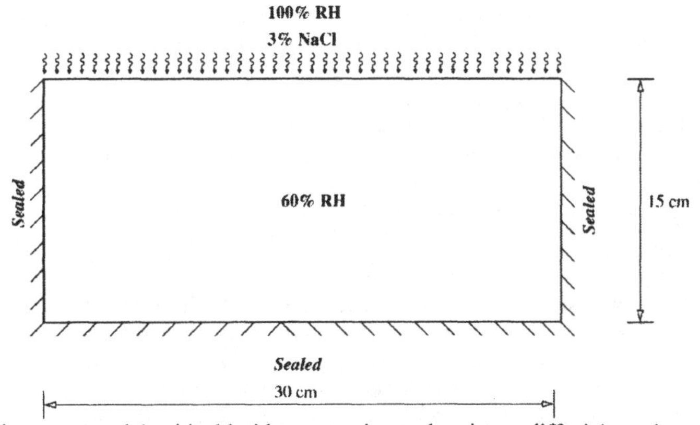

Fig. 7. The concrete slab with chloride penetration and moisture diffusion on the top surface

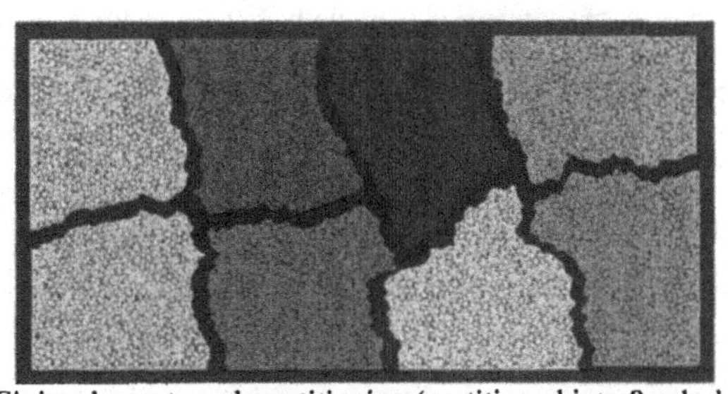

Fig. 8. Finite element mesh partitioning (partitioned into 8 sub-domains)

Fig. 8 shows the finite element mesh partitioned into 8 sub-meshes. The mesh is partitioned in such a way that each sub-mesh has almost the same number of elements. Fig. 9 shows the distribution of the chloride concentration after 550 days of ponding.

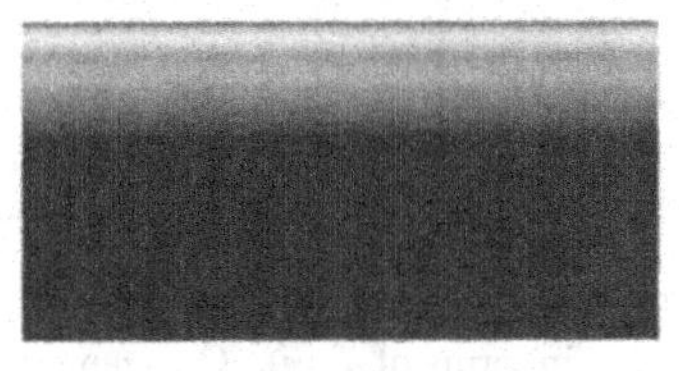

Fig. 9. Chloride ions distribution in the concrete slab after 550 days

For the same computational problem, different parallel programming algorithms and different partitioning of the structural domain lead to different efficiency, i.e. reduction of computing time. We used two performance parameters to evaluate the performance of the parallel implementation of finite element program. The two parameters are speed-up and efficiency. The speed-up is the ratio between the time taken by the code to execute on a single processor, $T_s(n)$, and the time taken for the same code to execute on p processors, $T_p(n)$,

$$S_p(n) = \frac{T_s(n)}{T_p(n)} \tag{19}$$

where n is the problem size (e.g. number of nodes in the finite element model). There are several different ways to interpret $T_s(n)$ in Eq. (19). One way is to consider $T_s(n)$ as the time taken to run the fastest serial algorithm on the fastest serial machine available, and another way is to take $T_s(n)$ as the time taken to run the fastest serial algorithm on the fastest processor in their parallel computing systems. The third possibility is to take $T_s(n)$ as the time taken to run the parallel algorithm on one processor. A higher value of S_p means a better parallel algorithm and partitioning. In our case, we replaced $T_s(n)$ with $T_2(n)$, the time taken to run the parallel algorithm on two processors. In addition to speed-up, parallel efficiency are often use for performance evaluation, which is the ratio between speed-up and the number of processors,

$$E_n(n) = \frac{S_p(n)}{} \tag{20}$$

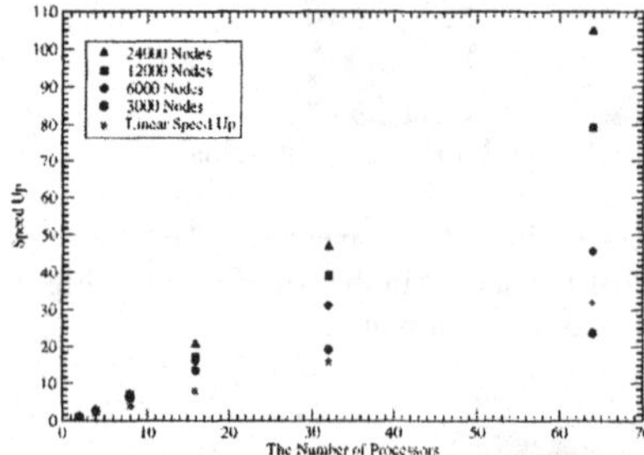

Fig. 10(a). Speed up over a number of processors using $T_2(n)$ as reference

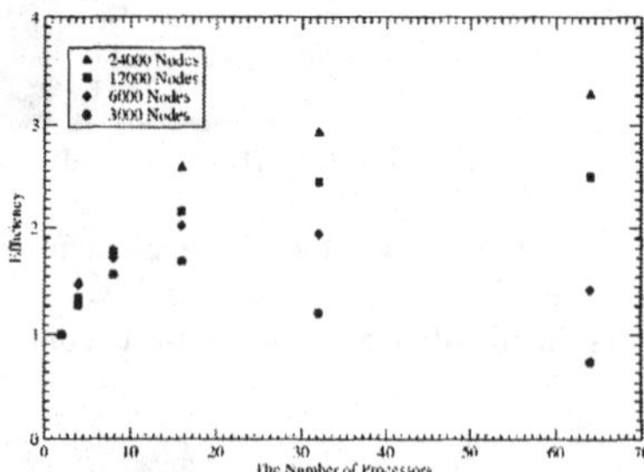

Fig. 10(b) Efficiency over a number of used processors using $T_2(n)$ as reference

Fig. 10(a) and Fig. 10(b) show the speed-up and efficiency by different mesh size, n, and different number of processors, p, in terms of $T_2(n)$. One can see that as p increases, the size of the mesh solved by each processor decreases. So, the speed-up increases with increasing p. This simply means that the time required for solving the problem becomes shorter when more processors are used in the analysis. But, with more processors, the interface boundaries between partitions increase, and the communication between the processors increases, and as a result, the total computing time increases. These are the reasons why there will be an optimal number of processors for a specific mesh size n beyond which increasing p will not improve the speed and efficiency of the computation. For example, the speed-up for the size of 3000 nodes reaches the optimal condition at 64 processors, see Fig. 10(a). This optimal number of processors is shown more clearly in Fig. 10(b). Another trend shown in the figures is that with increasing mesh size *n*, the efficiency increases at a fixed number of processors. This is because the communications between the sub-domains becomes relatively less. Overall, at 64 processors, the speed-up and efficiency for the largest problem size reach 107 and 3.3, respectively. This shows that the parallel algorithm performs quite well for the coupled chloride and moisture diffusion problem.

CONCLUSIONS

1) Moisture diffusion and chloride penetration are considered as two coupled driving forces responsible for the transport of moisture and chloride in concrete. Fick'law of chloride diffusion is modified to include the effect of moisture gradient, and Fick'law of moisture diffusion is modified to include the effect of chloride gradient. These modifications are similar to the modifications to Fourier law and Fick's law when the coupling effects between moisture transfer and heat conduction in porous media were considered (i.e. Soret flux and Dufour flux).
2) In the modified Fick's laws for moisture and chloride diffusions, there are two coupling parameters, D_{Cl-H} and D_{H-Cl}. D_{Cl-H} represent the effect of moisture diffusion on chloride penetration and D_{H-Cl} is for the effect of chloride penetration on moisture diffusion. New experimental methods are proposed to verify the significance of the two coupling effects.
3) Analysis methods are developed to evaluate the coupling parameter D_{Cl-H} and D_{H-Cl} based on the experimental results obtained from the proposed new testing methods.
4) The present test data show that the coupling effects between moisture diffusion and chloride penetration are very significant. The coupling terms in the governing partial differential equations cannot be simply neglected. As a result, the two governing equations are fully coupled.
5) The present test data show that both D_{Cl-H} and D_{H-Cl} are not constants but increase with chloride concentration. This is an important result, because the concentration dependence of the two coupling parameters makes the two governing equations not only fully coupled but also nonlinear, which is a great challenge for developing computational methods to solve these equations.
6) More systematic experimental studies are needed to investigate the two coupling parameters. Different concrete mix design parameters should be examined in the future studies, which is very important for developing material models for the two coupling parameters.
7) A parallel finite element program is developed to solve the two fully coupled nonlinear partial differential equations. The parallel algorithm is implemented using the functions of PETSc and MPI. The performance of the parallel finite element program is evaluated by two indicators: the speed-up and the efficiency.
8) With increasing number of processors, the required computing time decreases, but the internal communications between the sub-domains increase, which reduce the speed-up. Therefore, there exist an optimal number of processors for each specific application. Depending on the physical problem under consideration, especially on the number of finite elements used, the optimal number of processors varies.

ACKNOWLEDGEMENTS

Partial financial support under NSF grant CMS-9872379 to University of Colorado at Boulder is gratefully acknowledged. Partial financial support under NSF grant ACI-0112930 to University of Colorado at Boulder is gratefully acknowledged.

REFERENCES

Ababneh, A., Benboudjema, F., and Xi, Y. (2003). "Chloride Penetration in Nonsaturated Concrete." *Journal of Materials in Civil Engineering, ASCE*, 15(2), 183-191.

Ababneh, A. and Xi, Y. (2002). "An Experimental Study on the Effect of Chloride Penetration on Moisture Diffusion in Concrete." *Materials and Structures*, 35, 659-664.

Abarr, L. (2005) "The Effect of Moisture Diffusion on Chloride Penetration in Concrete", MS Thesis, University of Colorado at Boulder.

Andrade, C. and Sanjuan, M.A. (1994). "Experimental Procedure for the Calculation of Chloride Diffusion Coefficients in Concrete from Migration Tests." *Advances in Concrete Research*, 6(23), 127-134.

Saetta, A.V., Scotta, R.V., and Vitaliani, R.V. (1993). "Analysis of Chloride Diffusion into Partially Saturated Concrete." *ACI Materials Journal*, 90(5), 441-451.

Swaddiwudhinpong, S, Wong, S.F., Wee, T.H., and Lee, S.L. (2000). "Chloride Ingress in Partially Saturated and Fully Saturated Concretes." *Concrete Science and Engineering*, 2, 17-31.

Tang, L. and Nilsson, L.O. (1992). "Chloride Diffusivity in High Strength Concrete." *Nordic Concrete Research*, 11, 162-170.

Wee, T.H., Wong, S.F., Swadiwudhipong, S., and Lee, S.L. (1997). "A Prediction Method for Long-Term Chloride Concentration Profiles in Hardened Cement Matrix Materials." *ACI Materials Journal*, 94(6), 565-576.

Suwito, X.-C. Cai, and Y. Xi (2006) "Parallel Finite Element Method for Coupled Chloride Penetration and Moisture Diffusion in Concrete", *International Journal of Numerical Analysis and Modeling* (in press).

Xi, Y. (1995a). "A Model for Moisture Capacities of Composite Materials, Part I: Formulation." *Computational Materials Science*, 4, 65-77.

Xi, Y. (1995b). "A Model for Moisture Capacities of Composite Materials, Part II: Application to Concrete." *Computational Materials Science*, 4, 78-92.

Xi, Y. and Bazant, Z. (1999). "Modeling Chloride Penetration in Saturated Concrete." *Journal of Materials in Civil Engineering*, ASCE, 11(1), 58-65.

Xi, Y., Bazant, Z.P., Molina, L., and Jennings, H.M. (1994a). "Moisture Diffusion in Cementitious Materials: Adsorption Isotherm." *Advanced Cement Based Materials*, 1, 248-257.

Xi, Y., Bazant, Z.P., Molina, L., and Jennings, H.M. (1994b). "Moisture Diffusion in Cementitious Materials: Moisture Capactiy and Diffusivity." *Advanced Cement Based Materials*, 1, 258-266.

Xi, Y., Willam, K., and Frangopol, D. (2000). "Multiscale Modeling of Interactive Diffusion Process in Concrete." *Journal of Materials in Civil Engineering, ASCE*, 126(3), 258-265.

Xi, Y. and Nakhi, A. (2005) "Composite Damage Models for Diffusivity of Distressed Materials", *J. of Materials in Civil Engineering*, ASCE, May/June, 17(3), 286-295.

DIRECT PHASE-RESOLVED STRAIN MEASUREMENTS IN CEMENTITIOUS MATERIALS

J. J. Biernacki, S. E. Mikel, C. J. Parnham
Tennessee Technological University
Department of Chemical Engineering
Box 5013
Cookeville, TN 38505

R. Wang
Georgia Institute of Technology
Department of Mechanical Engineering
801 Ferst Dr. N. W.
Atlanta, GA 30332-0405

J. Bai
Materials Processing Center
University of Tennessee
Knoxville, TN 37996

T. Watkins, M. J. Lance and C. Hubbard
Oak Ridge National Laboratory
Oak Ridge Tennessee, TN 37831-6064

ABSTRACT

Most common mechanisms of concrete degradation involve the development of localized stresses and cracking on the micro-scale. Unfortunately, techniques for probing in situ micro-scale mechanical response are limited or underdeveloped for cementitious systems. The present paper, however, describes a diffraction-based and a luminescence-based technique for making direct, phase-resolved micro- and meso-scale strain measurements. Such methods are independent of externally applied strain sensors and instead utilize one or more of the native phases, or a suitable phase deliberately introduced, as an internal strain gauge.

Coherent scattering of x-rays results in diffraction. This interaction between photons (x-ray) and regularly arranged atoms within a crystal is sensitive enough that minute changes in the spacing of the atoms within a crystal can be directly measured, for example, due to applied loads. The present paper summarizes this technique and illustrates the method. Though x-rays can be used to make direct measurement of the distance between atoms in a crystalline solid, strains can be inferred from other properties as well. Luminescence spectroscopy, for example, is directly related to the stored bond energy and electronic configuration of the atoms in a crystal, both of which are strain sensitive. In particular, chromium (Cr^{+3}) doped aluminum oxide (A) exhibits a strong luminescence strain response that has been well characterized. Here, small amounts of tiny single crystals of Cr^{+3} doped A were seeded into portland cement paste and used as internal strain sensors. The response of an individual A seed particle, on the order of 10 μm in size, to externally applied stresses of up to 35 MPa are presented to illustrate the method.

INTRODUCTION

It is well accepted today that concrete durability is intimately linked with the transport properties of the material and the literature is extensive on the relationship between pore structure, chemistry and resulting modes of failure. A common thread linking most of these degradation mechanisms is the development of stresses that ultimately lead to cracking and deterioration of mechanical properties. Presently, the grand majority of test methods for interpreting the mechanical response, resulting from the interaction of transport properties and the environment, utilize macroscopic or mesoscopic indicators. For example, sulfate attack is typically described in terms of mortar bar elongation per the standard test method ASTM C 1260-05[1], a macroscopic indicator. And, while some studies utilize microscale imaging, most infer mechanical response from observed changes in microstructure rather than by directly measuring stress or strain[2]. Furthermore, only a limited number of techniques have been developed that have phase resolution, i.e. direct imaging[3, 4], speckle pattern interferometry combined with visible light imaging[5] and other related techniques.

A number of prior studies, however, point to various other micro and mesoscale methods that might be further developed, including, neutron diffraction[6], x-ray diffraction[7] and luminescence spectroscopy methods[8, 9, 10]. This paper describes two such techniques, ψ-tilting x-ray diffraction and photo stimulated luminescence spectroscopy (PSLS), documents their development and illustrates and reviews some early results.

EXPERIMENTAL

Sample Preparation

Samples of neat, type-I ordinary portland cement and mortars of various composition made from the same, see Table I, were prepared by hand mixing the cement, and aggregate where used, with water.

Table I. Sample formulations.

Sample Type	Water/Cement (w/c)	Aggregate Type	Aggregate/Cement
Neat Cement	0.4	None	0.0
Silica Mortar	0.4	Sigma Aldrich 99.99% SiO_2, hand ground	2.0
Alumina Mortar	0.4	AK 502, Sumitomo Chemicals	0.01

The mixtures were placed in vials of nominally 2.54 cm in diameter and 5 cm long. The cement was permitted to cure at 35 °C for at least one day after which time the vial (mold) was stripped and the sample placed under saturated lime water at the same temperature. Curing was continued for at least seven days, after which time the billet was cut into rectangular prisms of 10.16×10.16×15.24 mm (0.4×0.4×0.6 inch) to a tolerance of ±0.01 mm. The exact curing time and conditions were not of great importance in this experiment since no conclusions regarding curing were sought. Since samples were analyzed over a broad time period, some samples were effectively hydrating for many months in ambient conditions post cutting into prisms.

X-ray Measurements

X-ray diffraction data was collected at the X14A beam line at the National Synchrotron Light Source (NSLS) located at Brookhaven National Laboratory (BNL). The beam line is operated by the High Temperature Material Laboratory (HTML) which is part of Oak Ridge National Laboratory (ORNL). The beam line configuration is detailed elsewhere[11, 12]. Data was collected at an x-ray wavelength of 0.15467 nm. The x-ray spot size was nominally 2 mm×0.5 mm with the 2 mm axis oriented collinear with the loading vector. Figure I illustrates the beam path, goniometer and load frame configuration.

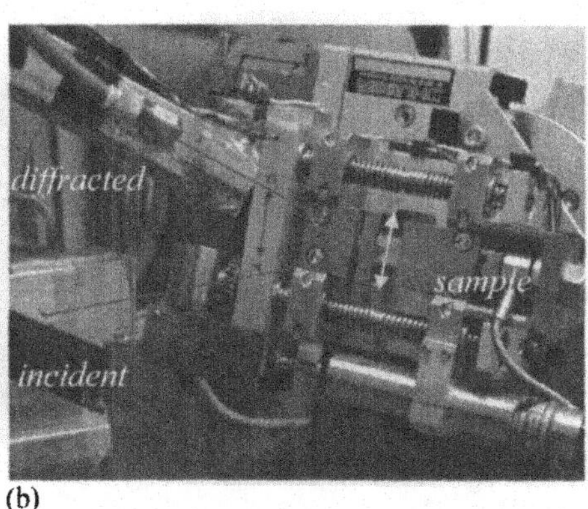

(a) (b)

Figure I. X-ray beam path, goniometer and load frame configuration at BNL beam line X14A. (a) Goniometer and load frame illustrating the laboratory direction (L_3), the sample direction S_3 and the tile angle ψ. (b) Illustration of beam path with sample and oscillation direction (perpendicular to the loading direction) shown as arrow superimposed above sample.

Samples were exposed to various no-load, load, no-load cycles using a computer controlled load frame and x-ray data collected for up to six ψ-tilt angles between 0.0 and 60°. No-load, load cycles varied between from 0.43 MPa (no-load) to as high as 34.5 MPa (load). Loading and unloading rates used were either instantaneous or 3.7 N/s (load removed in less than one second or 50 lb_f/min) depending upon the experiment. Specific load cycles and conditions are listed with the datasets for clarity, i.e. see Figure Vb. For neat portland cement paste, x-ray scans were collected so as to capture data for CH reflections for the (214)* plane, peak at nominally 131.1 °2θ. For silica mortar, the (413)† plane at nominally 139.8 °2θ was used.

Luminescence Spectroscopy Measurements

Luminescence spectra were obtained at the HTML using a Dilor XY800 triple stage Raman microprobe. The photon source consisted of a Innova 308C Ar^+ ion laser operating at 514.5 nm between 10 and 100 mW output power. The sample was mounted in compression using a computer controlled and automated load frame which was fixed to an electronically

* Data for the (312) plane is reported elsewhere[11].

† Data for the (206) and (330) planes are reported elsewhere[11].

controlled stage for precise adjustment of the position of the sample. Figure II illustrates the experimental configuration.

Luminescence spectra were collected for no-load, load, no-load cycles similar to that described above for x-ray experiments.

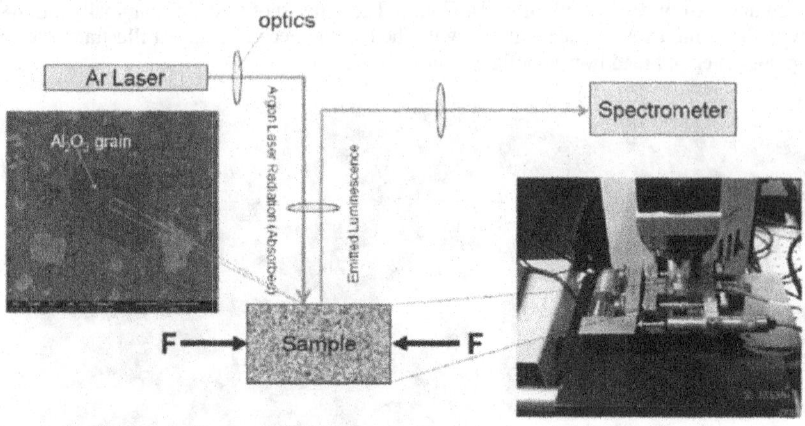

Figure 2. Luminescence spectroscopy set-up.

DISCUSSION

X-ray Diffraction-Based Strain Measurements

X-ray diffraction provides a direct measure of the distance between atomic planes in a crystalline lattice. Changes in the atomic spacing on the order of one part in 10^5 are now possible using highly parallel x-ray beams and sufficient counting statistics. Since the strain to failure in ordinary portland cement are on the order of 1 part in 10^3, such should be measurable using a diffraction based method. Furthermore, since diffraction provides detailed information regarding the spacing between atoms as a function of lattice direction, this method makes it possible to observe not only phase resolved, but also anisotropic responses.

The application of x-rays to the measurement of stress states is well developed for crystalline materials and to a lesser extent for composites[13]. Application of x-ray diffraction to cementitious systems was first illustrated by Shchukin, et al.[14], however, lack of statistical analysis, the use of a laboratory x-ray source and choice of diffraction angle places doubt in their results. Schulson and coworkers[6] have demonstrated a related method using neutron diffraction, however, were unable to measure strains due to lack of statistical confidence. More recently, Biernacki, et al.[11, 12] have presented preliminary results using synchrotron x-rays which suggest that an x-ray-based approach is possible. Details of the method used by Biernacki, et al. are given elsewhere[11, 12] and so are only summarized here.

X-ray diffraction is governed by Bragg's Law which relates the interplanar spacing, d, to the diffraction Bragg angle θ:

$$d = \frac{n\lambda}{2Sin\theta} \quad (1)$$

where n is an integer number and λ is the incident radiation wavelength. Upon differentiation, one sees that the changes in diffraction angle increase with increasing angle for constant strain ε:

$$\Delta\theta = -\varepsilon \tan\theta \quad (2)$$

This implies that better instrumental strain sensitivity is achieved when high angle (high 2θ) reflections are observed. In fact, it is generally suggested that this method not be applied to diffraction angles below about 130 $^o2\theta$.

For diffracting crystals, the orientation of the diffracting planes within the sample relative to the direction of the applied load is also important since the strain observed in the crystal transforms to strain observed in the sample via the following tensor relationship:

$$\langle \varepsilon'_{ij} \rangle = a_{ij} a_{jl} \langle \varepsilon_{kl} \rangle \quad (3)$$

utilizing the Einstein summation convention, where i, j, k, and l are the integer (each can be 1, 2, 3) corresponding to the directions of the orthoganol corrdinate system and a_{ik} and a_{jl} are the direction cosines. Here, uniaxial loading is applied in the "2" direction. The prime directions are the corresponding orientation of the diffracting crystallites relative to the unprimed directions and directions 1' and 3' define the θ-2θ plane with direction 3' being the normal to the diffracting plane and 1' lying in the plane of the diffracting atoms. Thus, the observed strain in any crystal rotated by an angle ϕ about direction "3" and ψ about direction "1" is given by:

$$\langle \varepsilon'_{33} \rangle = \langle \varepsilon_{11} \rangle \cos^2\phi \sin^2\psi + \langle \varepsilon_{12} \rangle \sin 2\phi \sin^2\psi + \langle \varepsilon_{22} \rangle \sin^2\phi \sin^2\psi + \langle \varepsilon_{33} \rangle \cos^2\psi + \langle \varepsilon_{13} \rangle \cos\phi \sin 2\psi + \langle \varepsilon_{23} \rangle \sin\phi \sin 2\psi \quad (4)$$

Hence, by tilting and rotating at various angles ϕ and ψ it is possible to observe the effective strain in crystallites at all possible orientations relative to the applied load. Experimentally, ϕ may be held constant at 0.0^o while ψ is varied over some range that is permissible by the goniometer and reflection being used, usually between about $\pm 75^o$. This method is generally referred to as ψ-tilting and provides the relationship between sample strain and crystallite strain. By applying the constitutive relationship for isotropic linear elastic materials to Equation (4) and then simplifying to a plane stress situation, the crystal strain may be expressed in terms of the sample stresses, giving the
expression shown in Equation (5).

$$\langle \varepsilon'_{33} \rangle = \frac{(1+\upsilon)}{E} \sigma_2 Sin^2\psi + \frac{\upsilon}{E} \sigma_2 \quad (5)$$

where, E is the elastic constant and σ_2 is the stress in the loading direction "2". This relationship implies that the crystal strain $\langle \varepsilon_{ij}' \rangle$ is a linear function of $\sin^2\psi$ when subject to these simplifying assumptions.

X-ray analysis of this sort relies on simultaneously gathering diffracted energy from a large number of crystallites at each tilt angle. A typical hydrated cement paste has a microstructure similar to that illustrated in Figure III with large unreacted clinker islands made up of one or a few large single crystals and a distribution of CH in small to large clusters. Many crystallites are much smaller in size than the typical cluster with CSH gel, other hydration products and porosity filling the remaining space.

When applying the ψ-tilting technique to a coarse grained, multiphase, polycrystalline material, such as portland cement paste or cement-based mortars, it is advisable to determine the randomness of crystallite orientation by obtaining a Debye-Scherer image. Here, a piece of Polaroid film was exposed by placing it in the backscattered beam path, see Figure IV.
The resulting rings are typical of a polycrystalline, fine grained material, while the sparsely distributed bright spots indicate coarse grains (large crystal size) and/or oriented crystals. While the ring spacings were not indexed, it would appear that the continuous diffraction rings are likely the result of scattering from CH crystallites which are present as randomly distributed fine grains (see Figure III). Although preferred orientation has been shown for CH near unreacted clinker surfaces[14], the bulk of the CH appears to be random enough and present is small enough crystallite size to produce more-or-less continuous rings. These rings indicate that peak intensities should be a week function of tilt angle ψ[12] and thus it should be possible to use CH as a strain sensor with the ψ-tilting approach.

The bright spots are likely due to large unreacted clinker cores. At any given ψ, angle a reflection may or may not be present due to the particle statistics associated with the coarse crystals. Though it may still be possible to extract strain data from such a coarse grained material at a particular tilt, it will be difficult to maintain signal intensity when the ψ angle is changed. The use of these phases remains largely unexplored at this time and will not be discussed further in this work.

Two series of tilt angles were collected for a no-load, load, no-load cycle of 0.43, 34.5, 0.43 MPa. Six tilt angles were selected[12] such that strong reflections were visible, see Figures Va and Vb. Although the Debye-Scherer data suggests that CH is finely distributed small grains, the non-linearity in Figure Va suggests some orientation and/or large grains are present. To further explore the effect of particle statistics on the observed strain vs. $Sin^2\psi$ relationship, an oscillating stage was installed so that a larger volume of particles could be sampled. Oscillation in the S_1 direction was chosen since it would not effect peak location. This change in experimental procedure was found to have a dramatic effect on the data. When using oscillation to improve particle counting statistics the highly non-linear data illustrated in Figure Va collapses to the nearly linear data illustrated in Figure Vb. Furthermore, the statistical confidence in the location of peak positions is greatly improved, making it possible to measure relatively small strains, of the order of 1/10,000, with reasonable accuracy.

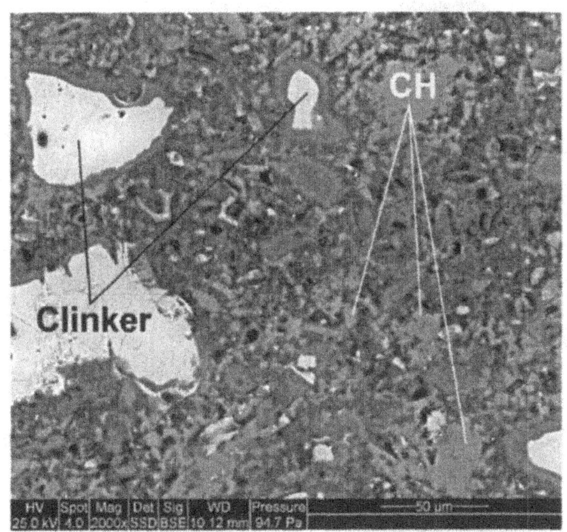

Figure III. Typical microstructure of a hydrated Portland cement paste.

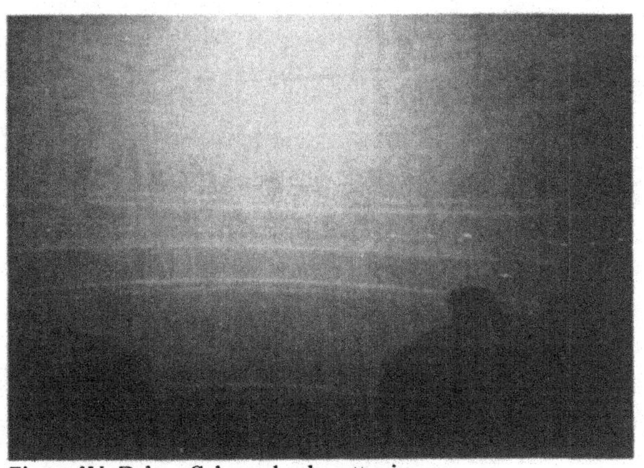
Figure IV. Debye-Scherer backscatter image.

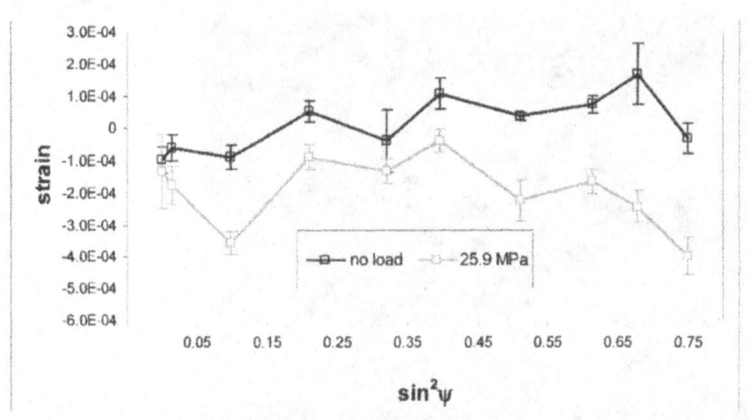

Figure Va. No-load (0.43 MPa), load (34.5 MPa), no-load (0.43 MPa) cycle without oscillation.

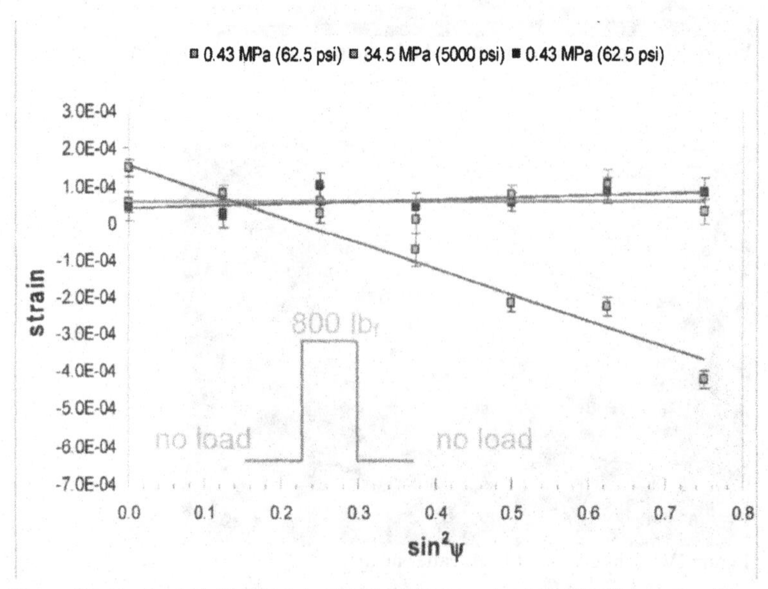

Figure Vb. No-load (0.43 MPa), load (34.5 MPa), no-load (0.43 MPa) cycle with oscillation.

Finally, this method may also be useful in the study of time dependent mechanical response, Figure VI. Comparisons were made between the strain in the SiO_2 phase of mortars prepared using pure crystalline silica and ordinary Potland Cement in response to various loading and unloading protocols. The objective in these experiments was to investigate any differences in strain state as loads were applied and removed at different rates. Sequential stress cycles were applied to the mortar specimens in the same manner as for the neat cement except that instead of removing the load at 3.7 N/s (50 lb_f/min), the load was removed instantaneously. Doing so gives the data shown below in Figures VIa through VIc. In Figure VIa the initial no load strain state data (the nominally flat data series between the two other series) was collected, then a 32.3 MPa stress (corresponding to 750 lb_f) was applied at 3.7 N/s. Under this load the strain state becomes compressive as shown by the negative slope of the regression curve. When the load was removed instantaneously, the strain state then changes to slightly tensile. Repeating the measurement after about 8 hours had elapsed showed that the strain state had changed from the slightly tensile state to a slightly compressive state very nearly identical to the initial strain state before the load was applied (Figure VIb). These data suggest that upon quickly removing the load, the crystal strain in the SiO_2 phase shows a time dependent behavior, initially changing rapidly before relaxing to the previous no load strain state. This is likely due to some type of strain relaxation, creep, or other time-dependent viscoelastic phenomenon of the mortar composite. When the sample was again loaded to 32.3 MPa with the load removed slowly at 3.7 N/s, the resulting strain measurement and its replicate taken 8 hours later show very close agreement (Figure VIc), indicating that the time-dependent behavior is not nearly as pronounced when the load is removed slowly. These data show that the strain state in the SiO_2 aggregate is not only related to the magnitude of the applied load, but also to the rate at which the load is applied and the amount of time that has passed since its application. This illustrates that the potential utility of the x-ray method is not confined to the static stress cycling experiments discussed earlier, showing that it may be useful for a range of mechanical analyses.

Photo Stimulated Luminescence Spectroscopy (PSLS)

Photo stimulated luminescence spectroscopy, also referred to as piezo spectroscopy, measures minute changes in the luminescence spectra caused by applied or residual stresses. The origin of the piezo luminescence phenomena is completely different than diffraction. Defined as any process wherein the emitted energy is different than the incident energy, luminescence includes, fluorescence, phosphorescence and triboluminescence. A piezo effect is simply a strain sensitive effect, i.e. piezo resistance is a change in resistance due to strain. Hence piezospectroscopy is a strain sensitive spectroscopic measurement and photo-stimulated luminescence spectroscopy (PSLS) is a form of piezospectroscopy that utilizes stain sensitive changes in the luminescence spectra. For PSLS to work, the material of interest must have a luminescence response that can be utilized. While many substances, including those present in hydrated portland cement, are luminescent, their luminescent response is not useful for strain measurement due to diffuse or weak signal. Furthermore, many pure materials either have useless luminescence or are not luminescent at all. Aluminum oxide, Al_2O_3 (A) falls into this category. Fortunately, small naturally occurring amounts of Cr^{+3} ion in A generates an extremely strong strain sensitive luminescence response which has been well characterized.

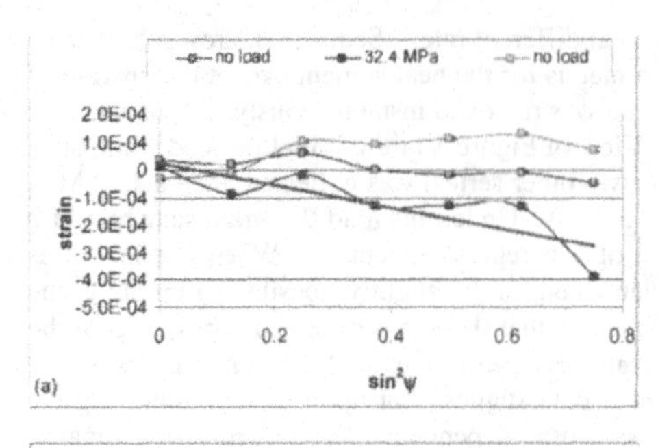

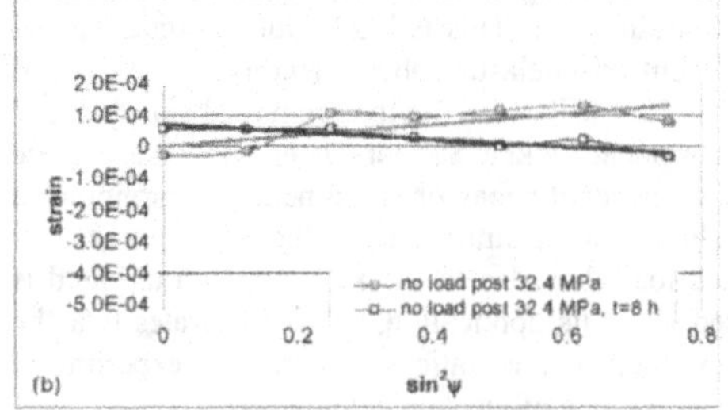

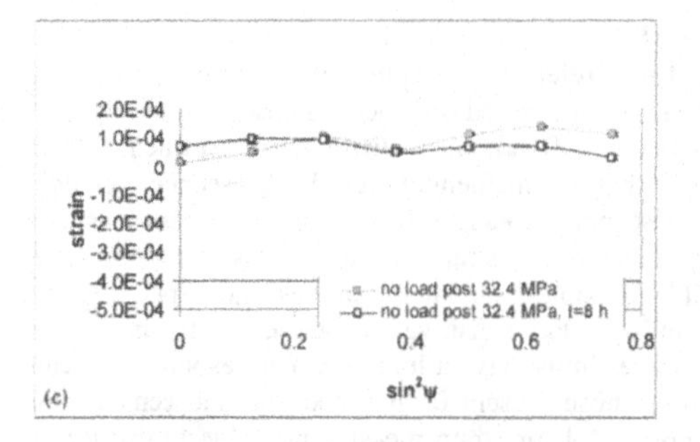

Figure VI. (a) Loaded at 3.7 N/s, unloaded instantaneously, post load data collected immediately after unloading. (b) Unloaded state from (a) compared to after allowing 8 h of rest. (c) Loaded at 3.7 N/s, unloaded at 3.7 N/s, post load data collected immediately after unloading and again after 8 h of rest.

When exposed to light of the correct wavelength, luminescent substances undergo an electron transition from the ground state to an excited state and back to ground. Upon transition from the excited to the ground state a photon is emitted. This photon is the luminescent response of the material. The electronic configuration of the luminescent species is defined by its orientation and its relative proximity and bonding relationship to it's nearest neighboring atoms in the lattice, in this case the Cr^{+3} ion in the aluminum oxide lattice. When strained, the shape and size of the electron orbitals change and hence, a small, but detectable change in luminescent emissions, in some cases, is quantifiable.

Alumina exhibits a pair of strong luminescence peaks centered at 14,430 and 14,400 cm^{-1} respectively, see Figure VII. The shift in frequency ν with an applied stress is given by:

$$\Delta \nu = \Pi_{ij} a_{ik} a_{jl} \sigma_{kl} \tag{5}$$

where Π_{ij} is the matrix of piezospectroscopic coefficients and σ_{kl} is the stress tensor.

Asmus and Pezzotti[10] describes the use of Cr^{+3} doped A as a strain sensor for the study of microscale mechanical response in cement paste. In this only know work in which piezospectroscopy was used to study the mechanical behavior of a portland cement-based material, Asmus and Pezzotti used a mortar prepared by mixing cement and submicron alumina particles (0.1 mm nominal particle size) in proportions of 4:1 (cement:A) with a w/c ratio of .4. Their laser spot size was 1 μm and they scanned an area of 40×70 μm. They concluded that the method is a useful tool for exploring micromechanics, however, they did not offer an error analysis making it difficult to apply a confidence interval to their results. Furthermore, including large amounts of extremely fine particles that are significantly stiffer than the matrix and than typical aggregate (370 GPa for Al_20_3 compared to 70 GPa for SiO_2), perturbs the system and may make mechanical extrapolation to other systems difficult.

The results of two very preliminary experiments are reported here. Luminescence spectra were collected for a single A particle over a period of about 40 minutes during which time the sample was loaded at a rate of 5.76 kPa/s (50 lb_f/min) from 44.6 N to 3564.6 N (10 lb_f to 800 lb_f), then held at 3564.6 N (800 lb_f) for 10 minutes after which time it was unloaded at -5.76 kPa/s (-50 lb_f/min) back to 44.6 N (10 lb_f). 35 individual spectra were collected at nominally equal intervals of about one minute. Error bars represent one standard deviation of the expected mean based on 60 datasets collected under no-load conditions. The results indicate that strains are being measured, however, the statistical confidence in the quantitative results at this time are somewhat low. Improved counting statistics and experimental procedures will likely reduce the observed variability.

In a follow-up experiment, the sample was cycled between no-load and load conditions and spectra were collected at each new stress state, e.g. at the no-load and load conditions. A plot of the observed stress in the A particle was plotted as a function of loading cycle for both the no-load states and the load states, resulting in two lines, see Figure VIII. The load state clearly becomes more compressive during load states of the successive cycles, indicating that the load state is being transferred to and indicated via the A particle. Not surprising, the no-load state also becomes more compressive with each successive cycle. This is likely the response of the A particle to the viscoelastic behavior of the cement matrix. Furthermore, initial loaded stress states appear to become more compressive relative to no-load states, a somewhat surprising response. This anomaly, while needing further investigation, is likely due to a non-uniform stress field on the micro-scale caused by material heterogeneity; inclusive of multiphase heterogeneities, porosity and extensive micro-cracking of the sample.

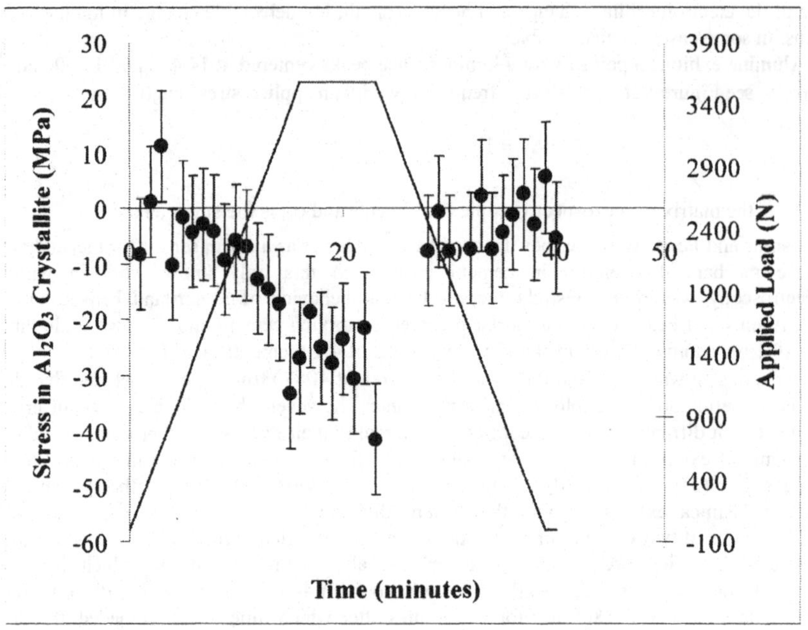

Figure VII. Stress response in Al_2O_3 due to gradual applied load‡.

CONCLUSION

Two methods that have not yet been exploited for use on portland cement-based materials have been explored and developed for measurement of direct micro- and meso-scale mechanical response: (1) ψ-tilting x-ray diffraction and (2) photo stimulated luminescence spectroscopy. The ψ-tilting method, previously described by Biernacki, et al., has now been developed to the point at which it can reliably be used to gather meso-scale ensemble averaged strain data for cement paste and mortar samples with aggregate particle sizes on the order of 10-50 μm. When using synchrotron x-rays to make measurement, statistically confident strains of 1/10,000 can be measured. The technique, however, requires knowledge of the orientation and distribution of particles that can easily be interpreted by combining simple SEM and Debye-Scherer backscattered x-ray images.

‡ An unfortunate programming error lost some data during the unloading cycle and is responsible for the gap in data shown on Figure VII.

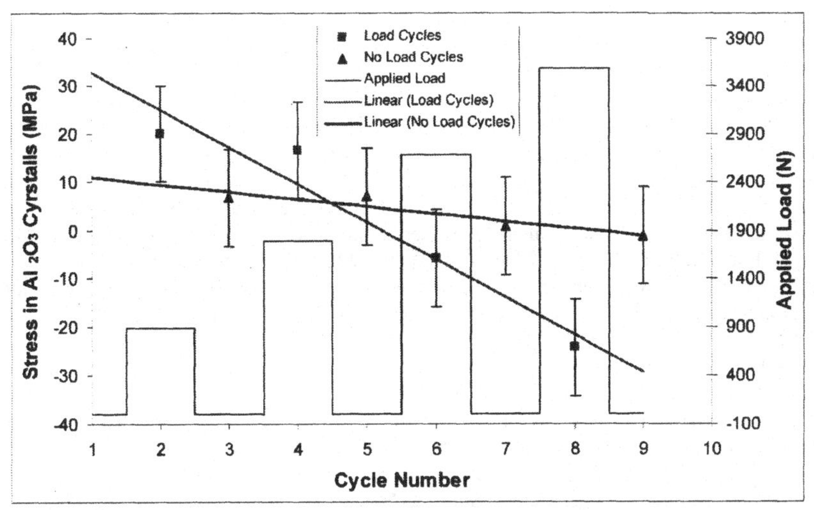

Figure VIII. Stress response in Al_2O_3 due to applied cyclic loading.

Photo stimulated luminescence spectroscopy has been demonstrated for the first time on a single micron-sized particle imbedded in a hydrated portland cement matrix. Small amounts of Cr^{+3} ion doped A particles, about 1% by mass of cement, can easily be observed as individual, unagglomerated entities within the matrix. At this time, data is very preliminary, but even early results indicate statistically measurable changes in stress (strain) states.

Additional research on these and related meso- and micro-scale techniques should realize new methods for making direct phase resolved mechanical response in cement-based materials. Such will likely be useful to the research community to develop multi-scale description of chemo-mechanical and physio-mechanical interactions that link transport related degradation mechanisms to mechanical failure modes that result in cracking cascades.

ACKNOWLEDGEMENTS

This material is based upon work supported by the National Science Foundation (NSF)[§] under Grant No. 0324616 and is partially supported by the Center for Manufacturing Research at Tennessee Technological University. This research was performed in part at the Oak Ridge National Laboratory, beamline X-14A at the National Synchrotron Light Source, Brookhaven National Laboratory and at the High Temperature Materials Laboratory, Oak Ridge National Laboratory, both sponsored by the Assistant Secretary for Energy Efficiency and Renewable Energy, Office of Transportation Technologies, as part of the High Temperature Materials Laboratory User Program, Oak Ridge National Laboratory, managed by UT-Battelle, LLC, for the U.S. Dept. of Energy under contract No. DE-AC05-00OR22725.

[§]Any opinions, findings, and conclusions or recommendations expressed in this material are those of the authors(s) and do not necessarily reflect the views of the National Science Foundation.

REFERENCES

1. ASTM C 1260-05, Standard Test Method for Potential Alkali Reactivity of Aggregate (Mortar-Bar Method), ASTM International (2005).
2. something image-based
3. Y. S. Choi and S. P. Shah, Measurement of Deformations on Concrete Subjected to Compression Using Image Correlation, Exp. Mech. 37 (3), 307-313 (1997).
4. C. M. Neubauer, H. M. Jennings and E. J. Garboczi, "Mapping Drying Shrinkage Deformation in Cement-Based Materials," Cem. Concr. Res., 27 (10), 1603-1612 (1997).
5. Z. Jia and S. P. Shah, Two-Dimensional Electronic Speckle Pattern Interferometry and Concrete Fracture Processes, Exp. Mech., 34 (3), 262-270 (1994).
6. E. M. Schulson, I. P. Swainson and T. M. Holden, "Internal Stress Within Hardened Cement Paste Induced Through Thermal Mismatch Calcium Hydroxide versus Calcium Silicate Hydrate," Cem. Concr. Res., 31, 1785-1791 (2001).
7. E.D. Shchukin et al., "X-ray Diffraction Method for the Determination of Residual Stresses in Cement," Kolloidnyi Zhurnal, 59, 96-101 (1997).
8. Q. Ma, D.R. Clarke, Stress Measurement in Single-Crystal and Polycrystalline Ceramics Using Their Optical Fluorescence, J. Am. Ceram. Soc. 76, 1433-1440 (1993).
9. Q. Ma, D.R. Clarke, Piezospectroscopic Determination of Residual Stresses in Polycrystalline Alumina, J. Am. Ceram. Soc. 77, 298-302 (1994).
10. S. M. F. Asmus and G. Pezzotti, Analysis of Microstresses in Cement Paste by Fluorescence Piezospectroscopy, Phy. Rev. E, 66, 052301 1-4 (2002).
11. Biernacki, J. J, C. Parnham, J. Bai, T. Watkins and C. Hubbard, "Meso-Scale Starin Measurements Using Synchrotron X-Rays," High-Performance Cement-Based Concrete Composites, J. J. Biernacki, S. P. Shah, N. Lakshmanan and S. Gopalakrishnan, eds., Am. Ceramic Soc., 291-99 (2005).
12. J. J. Biernacki, C. J. Parnham, T. R. Watkins, C. R. Hubbard, J. Bai, Phase-Resolved Strain Measurements in Hydrated Ordinary Portland Cement Using Synchrotron X-rays, J. Am. Cer. Soc., submitted.
13. I. C. Noyan and J. B. Cohen, "Residual Stress Measurement by Diffraction and Interpretation," Springer-Verlag, New York, pp.276 (1987).
14. R.J. Detwiler and P.J.M. Montiero, "Texture of Calcium Hydroxide Near the Cement-Paste Aggregate Interface,' Cem. and Concr. Res. 18, 823-829 (1988).

Modelling

CALCIUM SILICATE HYDRATE (C-S-H) SOLID SOLUTION MODEL APPLIED TO CEMENT DEGRADATION USING THE CONTINUUM REACTIVE TRANSPORT MODEL FLOTRAN

J. W. Carey and P. C. Lichtner
Los Alamos National Laboratory
Los Alamos, New Mexico

ABSTRACT

Reactive transport calculations have the potential to provide predictions of the progress and consequences of cement degradation as a function of position in the cement body. In the past, these calculations have been limited to describing the behavior of pure substances. In this work, a new approach for incorporating solid solutions into equations describing reactive transport in porous materials is applied to the C-S-H phase and to modeling cement degradation.

A thermodynamic model for C-S-H that spans the full range of potential compositions within the $Ca(OH)_2$–SiO_2 binary was developed from the data of Chen et al.[1] This model is used to reproduce experimental observations of decalcification of C-S-H as a function of decreasing pH. The solid solution model for C-S-H was incorporated into the reactive transport code FLOTRAN and applied to carbonation and sulfate-attack degradation mechanisms.

The reactive transport calculations successfully reproduce carbonation reaction features that were observed in a core of wellbore cement recovered from the SACROC Unit, a CO_2-enhanced oil recovery field in West Texas. These observations include the development of a 1-cm thick zone of calcium carbonate and amorphous silica and alumina phases separated from relatively unaltered cement by a narrow deposition zone. The simulation of sulfate attack reproduced the characteristic progressive transformation of monosulfate to ettringite. In both the carbonation and sulfate attack calculations, the model predicts progressive decalcification of C-S-H. Aqueous species-dependent diffusion coefficients were required to reproduce the observed and theoretical distribution of gypsum. The width and phase distribution of the reaction zones were sensitive to porosity, tortuosity, and mineral reaction rates.

INTRODUCTION

There are a number of significant obstacles to the development of accurate numerical models of the performance of Portland cement. In particular, compositionally variable and possibly metastable phases complicate efforts to represent the response of cement paste to changes in the chemical environment. Important chemical degradation mechanisms such as carbonation and sulfate attack cannot be fully understood without accounting for the compositional evolution of the primary cement paste phases.

A number of such compositional models exist for the C-S-H phase[2,3,4,5]. While these have been used in some studies of degradation processes[6,7], they have not been used in reactive transport calculations that form the basis of calculations of the depth and progress of chemical reactions in a cement body.

In the following, we describe recent developments in reactive transport modeling that allows calculation of the effect of compositional variation (solid solution behavior) on cement degradation processes. We begin by deriving a solid solution model for C-S-H that is capable of representing highly decalcified compositions. We then describe the features and methods of the reactive transport code, FLOTRAN. We conclude with example calculations of cement degradation by carbonation and sulfate attack.

C-S-H SOLID SOLUTION

Decalcification of the C-S-H phase is one of the key features of many cement degradation processes:

$$\mathrm{Ca}_r\mathrm{SiO}_{2+r}\cdot r\mathrm{H_2O} \longrightarrow \mathrm{Ca}_{r-y}\mathrm{SiO}_{2+r-y}\cdot(r-y)\mathrm{H_2O} + y\mathrm{Ca}^{2+} + 2y\mathrm{OH}^-. \tag{1}$$

This reaction occurs during carbonation, sulfate and acid attack, etc. Instead of a fixed composition having a limited range of stability, the solid solution behavior of C-S-H allows the phase to persist in a changing chemical environment. To the extent that decalcified C-S-H can maintain its cementing qualities, Equation 1 shows how a cement system can retain its structural integrity during some, perhaps limited, amount of chemical attack.

The development of a thermodynamic model for solid solution in C-S-H is thus a potentially important aspect of modeling cement degradation processes. There are a number of thermodynamic models in the literature[2,3,4,5,8] that could be adapted for use in reactive transport calculations. However, as we require both a simple analytical form for the thermodynamic properties and the possibility for representing the full range of compositional variation in C-S-H, we describe an approach for extracting thermodynamic properties of the solid solution from the solubility data of Chen et al.[1].

Thermodynamic Model for C-S-H

The objectives of our thermodynamic treatment of C-S-H are to predict the composition of the coexisting aqueous solution and C-S-H phase with sufficient accuracy to reproduce continuous variation of CSH composition according to Equation 1. In the present work, we do not attempt to reproduce the important differences in compositional trends of C-S-H as reflected in different structural C-S-H types[1,9]. Rather, we attempt to describe an average behavior of the C-S-H phase. In our approach, we were guided by Gartner and Jennings[8] who demonstrated that variation in C-S-H composition could be characterized by an exchange between idealized portlandite and amorphous silica end-members:

$$\begin{aligned}\mathrm{Ca}_r\mathrm{SiO}_{2+r}\cdot r\mathrm{H_2O} &\rightleftharpoons r\mathrm{Ca(OH)_2} + \mathrm{SiO_2(am)},\\ &\rightleftharpoons (1+r)\big[(1-x)\mathrm{Ca(OH)_2} + x\mathrm{SiO_2(am)}\big],\end{aligned} \tag{2}$$

where r is the ratio Ca/Si and x is the mole fraction of $\mathrm{SiO_2}$ in CSH which is equivalent to $1/(1+r)$.

Gartner and Jennings also showed that the chemical potential of the exchange reaction represented by Equation 2 showed a systematic variation as a function of CSH composition. However, they did not develop their approach into an analytical model. In a subsequent study, Jennings and his co-workers further elucidated the complex phase features of C-S-H in a thorough study of CSH and coexisting aqueous solutions [1]. We have used these data in the development of our CSH model.

In order to interpret CSH equilibria, we have used the Lippmann variable approach to solid-solution aqueous-solubility equilibria described by Glynn and Reardon [10] and Glynn et al. [11]. Kersten [4] also used this method for C-S-H but he limited the compositional range of analysis to $r > 1$. The Lippman method allows the simultaneous depiction and analysis of solution and solid compositions in terms of a total concentration (Lippman) variable. Dissolution of the C-S-H phase, written in terms of the mol-fraction variable x, is

$$Ca_{1-x}Si_xO_{1+x} \cdot (1-x)H_2O \longrightarrow (1-x)Ca^{2+} + 2(1-x)OH^- + xSiO_2(aq). \quad (3)$$

The Lippman variable (Σa) corresponding to Equation 3 is

$$\log(\Sigma a) = \log\left[(a_{Ca^{2+}})(a^2_{OH^-}) + a_{SiO_2(aq)}\right], \quad (4)$$

obtained from the sum of the ion activity products for the two endmembers.

Chen et al. [1] measured the solubility of six series of synthetic C-S-H phases which are shown in Figure 1, plotted in terms of the Lippman variable. The first three series were obtained from a C_3S starting material and the second three series were created by double decomposition of $CaNO_3$ and Na_2SiO_3. Chen et al. equilibrated these samples at various mass/water ratios and by differing decalcification treatments.

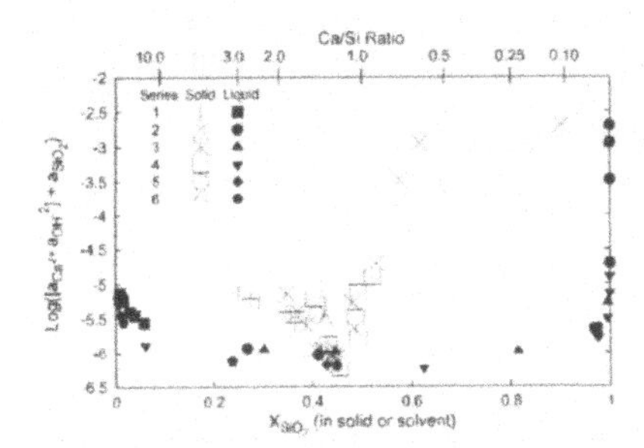

Figure 1: Solubility data for C-S-H from Chen et al. [1] plotted on a Lippman Diagram. Six experimental series are shown (series 1-3 were synthesized from C_3S and series 4-6 were synthesized by double decomposition of $Ca(NO_3)_2$ and $NaSiO_3$).

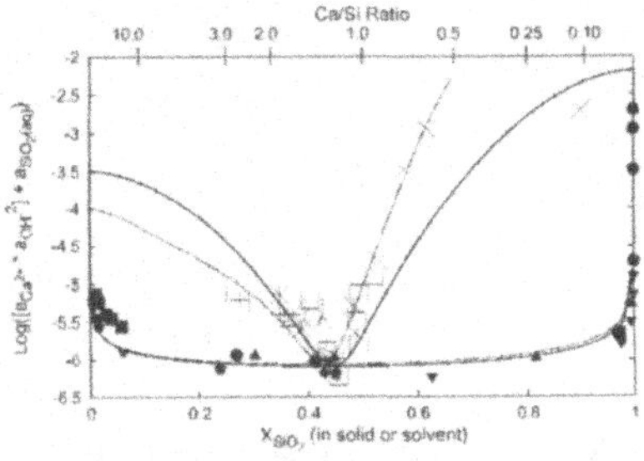

Figure 2: Regression results for a 5 parameter (solid line solidus and solutus) and 6 parameter (dotted line solidus and solutus) fit to the experimental data of Chen et al. [1]. See Figure 1 for meaning of symbols.

The Chen et al. data span a wide range of X_{SiO_2} (from 0.25 to 0.90 or in terms of Ca/Si, from 3.0 to 0.11). As shown in Figure 1, the CSH system exhibits a change in fluid/solid partitioning at $X_{SiO_2} \approx 0.44$: at smaller values the fluid is silica-depleted and at larger values the fluid is silica-enriched compared to the solid. At this crossover composition (alyotropic minimum), a solid dissolves congruently to a liquid of its own composition and a liquid precipitates a homogeneous solid of the same composition. At all other points, dissolution and precipitation processes are incongruent.

As might be expected, the solid-liquid equilibria in Figure 1 reflect non-ideal mixing behavior between the hypothetical $Ca(OH)_2$ and SiO_2 endmembers of the C-S-H system. For example, Glynn and Reardon[10] show that systems with an alyotropic minimum must have a negative Gibbs free energy of mixing (i.e., the endmembers combine favorably).

Following the approach of Glynn et al.[11], the non-ideal mixing properties of the C-S-H system were analyzed with a non-ideal asymmetric mixing model represented by the following excess Gibbs Free energy function:

$$G^E = x_1 x_2 RT\left[a_o + a_1(x_1 - x_2) + a_2(x_1 - x_2)^2 + a_3(x_1 - x_2)^3 + \cdots\right]. \tag{5}$$

This system was solved by independently fitting the experimentally measured Lippman variable ($\Sigma\alpha$, Equation 4) as a function of the values of $X^{CSH}_{SiO_2}$ in the solid and $X^{liq}_{SiO_2}$ in the liquid as reported by[1]. The equations for the solid and liquid are respectively[10]:

$$\log(\Sigma\alpha) = \log\left[K_{Ca(OH)_2}(1 - X^{CSH}_{SiO_2})\lambda_{(1-X^{CSH}_{SiO_2})} + K_{SiO_2}X^{CSH}_{SiO_2}\lambda_{(X^{CSH}_{SiO_2})}\right], \tag{6}$$

and

$$\log(\Sigma\alpha) = -\log\left[\frac{1 - X^{liq}_{SiO_2}}{K_{Ca(OH)_2}\lambda_{(1-X^{CSH}_{SiO_2})}} + \frac{X^{liq}_{SiO_2}}{K_{SiO_2}\lambda_{(X^{CSH}_{SiO_2})}}\right], \tag{7}$$

where $K_{Ca(OH)_2}$ and K_{SiO_2} are the endmember solubility constants and the activity coefficients, λ, are given by

$$\begin{aligned} RT\ln\lambda_{(X^{CSH}_{SiO_2})} &= (1 - X^{CSH}_{SiO_2})^2\left[a_0 + a_1(4X^{CSH}_{SiO_2} - 1) + (2X^{CSH}_{SiO_2} - 1)(a_2(6X^{CSH}_{SiO_2} - 1) + \right. \\ &\quad \left. a_3(16(X^{CSH}_{SiO_2})^2 - 10X^{CSH}_{SiO_2} + 1))\right], \end{aligned} \tag{8}$$

and

$$\begin{aligned} RT\ln\lambda_{(1-X^{CSH}_{SiO_2})} &= (X^{CSH}_{SiO_2})^2[a_0 + a_1(4X^{CSH}_{SiO_2} - 3) + (2X^{CSH}_{SiO_2} - 1)(a_2(6X^{CSH}_{SiO_2} - 5) + \\ &\quad a_3(16(X^{CSH}_{SiO_2})^2 - 22X^{CSH}_{SiO_2} + 7))]. \end{aligned} \tag{9}$$

The Chen et al. dataset includes 35 measures of aqueous Ca and SiO_2 concentration as well as the Ca/Si ratio of the coexisting C-S-H. They provide pH data for 15 of the 35 measurements. The system of equations to be solved (Equations 4 and 6 to 9) requires activities for Ca^{2+}, OH^-, and SiO_2. These were obtained by speciating the total concentrations of Ca and SiO_2 with the computer program *Geochemist's Workbench*^{TM} (Table 1).

The computer code PEST[12], which uses a model-independent parameter estimation method, was used to conduct the non-linear regression analysis for the unknowns $\mu^o_{Ca(OH)_2}$, $\mu^o_{SiO_2}$, a_0, a_1, a_2, and a_3 in Equations 6-9. Analyses with evenly weighted data were not capable of reproducing the lowest solubility values (near $X_{SiO_2} = 0.44$) because most of the aqueous data are located at either the extreme $Ca(OH)_2$-rich or SiO_2-rich endmembers. Thus some of the intermediate composition data were given more weight in the regression analysis. Models with 5 and 6 parameters are compared in Figure 2 with the resulting parameter estimates in Table 2.

Table 1: Experimental data obtained by Chen et al.[1] (Ca/Si, pH meas, Ca_T, and SiO_{2T}) together with speciation calculations obtained from *Geochemist's Workbench™* (OH^-, Ca^{2+}, and $SiO_2(aq)$ are in molality; na = not available).

Sample	Ca/Si	pH meas	Ca_T	SiO_{2T}	pH calc	OH^-	$\log a_{OH^-}$	Ca^{2+}	$\log a_{Ca^{2+}}$	$SiO_2(aq)$	$\log a_{SiO_2(aq)}$
1a	3.00	na	20.83	48.9	12.477	0.03747	-1.5180	0.0208	-2.1013	4.89e-005	-7.1163
1b	2.99	na	20.59	41.5	12.473	0.03708	-1.5222	0.0206	-2.1045	4.15e-005	-7.1811
1c	2.64	na	19.22	47.5	12.447	0.03475	-1.5483	0.0192	-2.1234	4.75e-005	-7.0832
1d	2.42	12.49	18.59	45.7	12.434	0.03367	-1.5608	0.0186	-2.1327	4.57e-005	-7.0812
1e	1.87	12.43	15.86	51.1	12.374	0.02898	-1.6212	0.0159	-2.1774	5.11e-005	-6.9447
1f	1.81	na	15.22	59.3	12.358	0.02786	-1.6371	0.0152	-2.1891	5.93e-005	-6.8575
1g	1.75	na	13.57	59.3	12.314	0.02498	-1.6810	0.0136	-2.2223	5.93e-005	-6.7959
1h	1.05	11.69	3.17	150.7	11.730	0.006009	-2.2649	0.00317	-2.6876	0.000151	-5.6787
2a	1.40	12.21	8.59	66.4	12.136	0.01612	-1.8586	0.00859	-2.3597	6.64e-005	-6.5119
2b	1.27	12.12	7.20	66.4	12.067	0.01360	-1.9282	0.00720	-2.4150	6.64e-005	-6.4251
2c	1.06	11.79	3.37	164.3	11.755	0.006375	-2.2405	0.00337	-2.6664	0.000164	-5.6683
2d	0.89	11.16	1.73	635.3	11.405	0.002777	-2.5900	0.00173	-2.9023	0.000635	-4.7098
2e	0.74	10.13	1.27	2303.6	10.702	0.0005451	-3.2928	0.00127	-3.0111	0.00230	-3.4864
2f	0.62	9.97	0.95	2890.7	10.118	0.0001406	-3.8775	0.000950	-3.1218	0.00289	-2.9508
2g	0.11	9.53	0.30	2565.7	9.394	2.595e-005	-4.6007	0.000300	-3.5810	0.00257	-2.7018
3a	1.08	11.63	2.37	271.1	11.596	0.004363	-2.3988	0.00237	-2.7897	0.000271	-5.2787
3b	1.22	11.97	4.81	93.6	11.905	0.009155	-2.0906	0.00481	-2.5460	9.36e-005	-6.0824
3c	1.33	12.15	7.36	82.1	12.075	0.01387	-1.9200	0.00736	-2.4080	8.21e-005	-6.3430
3d	1.45	12.10	8.31	69.3	12.123	0.01561	-1.8717	0.00831	-2.3700	6.93e-005	-6.4768
3e	1.58	12.25	13.08	52.5	12.300	0.02413	-1.6950	0.0131	-2.2330	5.25e-005	-6.8295

Sample	Ca/Si	pH meas	Ca_T	SiO_{2T}	pH calc	OH^-	$\log a_{OH^-}$	Ca^{2+}	$\log a_{Ca^{2+}}$	$SiO_2(aq)$	$\log a_{SiO_2(aq)}$
3f	1.70	na	15.90	41.1	12.375	0.02906	-1.6201	0.0159	-2.1767	4.11e-005	-7.0409
3g	1.87	12.46	20.59	41.5	12.473	0.03708	-1.5222	0.0206	-2.1045	4.15e-005	-7.1811
4a	1.48	12.37	19.27	16.8	12.448	0.03487	-1.5468	0.0193	-2.1228	1.68e-005	-7.5366
4c	1.44	12.16	17.82	15.9	12.419	0.03239	-1.5764	0.0178	-2.1445	1.59e-005	-7.5167
4d	1.31	11.88	9.96	20.4	12.196	0.01864	-1.7995	0.00996	-2.3144	2.04e-005	-7.1001
4e	1.20	11.70	5.08	44.1	11.929	0.009710	-2.0662	0.00508	-2.5280	4.41e-005	-6.4375
4g	1.03	11.46	2.70	97.3	11.668	0.005169	-2.3274	0.00270	-2.7440	9.73e-005	-5.8000
4h	1.05	11.56	2.03	144.0	11.543	0.003837	-2.4520	0.00203	-2.8458	0.000144	-5.4962
4i	0.92	11.16	1.66	421.6	11.417	0.002850	-2.5780	0.00166	-2.9182	0.000422	-4.8992
4j	1.03	na	1.41	220.4	11.373	0.002563	-2.6217	0.00141	-2.9794	0.000220	-5.1349
5a	1.45	na	15.37	15.7	12.363	0.02817	-1.6325	0.0154	-2.1864	1.57e-005	-7.4410
5b	1.26	na	6.25	42.5	12.012	0.01188	-1.9834	0.00625	-2.4603	4.25e-005	-6.5517
6a	1.35	12.33	13.85	16.6	12.323	0.02552	-1.6723	0.0138	-2.2164	1.66e-005	-7.3608
6b	1.28	12.07	7.37	32.4	12.077	0.01394	-1.9178	0.00737	-2.4077	3.24e-005	-6.7494
6c	1.26	12.03	6.09	41.9	12.001	0.01158	-1.9937	0.00609	-2.4687	4.19e-005	-6.5455

Table 2: Thermodynamic parameters for the C-S-H system obtained by regression of the Chen et al.[1] data.

Parameters	5 Parameter Model	6 Parameter Model
$\mu^{o}_{Ca(OH)_2}$	-3.49	-4.0
$\mu^{o}_{SiO_2}$	-2.19	-2.6
a_0	-29.67	-24.67
a_1	0.28	5.0
a_2	-0.0032	18.0
a_3	–	20.0

A comparison of the C-S-H model results with the data of Chen et al.[1] as well as the models of Reardon[3] and Kersten[4] is given in Figures 3 and 4. These show that the model represents a compromise between two or three distinct trends in the experimental data. Because the model was constructed with calcium hydroxide and silica endmembers, it is capable of providing predictions for the full range of experimental data, including low Ca/Si–low pH C-S-H phases.

All of the models show the correct behavior at the high pH limit (portlandite solubility) with predictions of asymptotically increasing Ca/Si ratios (Figure 3). At low pH, the model developed here predicts continuous decalcification and splits the difference between experimental data in the range of $9.5 < \mathrm{pH} < 10.2$. The models of Reardon and Kersten are limited to Ca/Si ≥ 1 and provide predicted compositions at pH > 10.5 or 11.5.

The models show considerable differences in compositional trends as a function of Ca concentration (Figure 4). Reardon's model follows the low Ca/Si compositional trend; Kersten's model reproduces the asymptotically increasing compositional trend; and the model developed here is intermediate.

All three models are in general agreement for Ca/Si ratios between about 1.1 and 1.5. The differences in the models are partly based on differences in the data sets used to constrain the model behavior and on differences in which features the authors were most interested in modeling. Any of these models are readily adapted to the reactive transport modeling approach described in this paper. Indeed, in as much as the C-S-H in Portland cement paste consists of multiple phases, some combination of several C-S-H phases with distinct compositional trends may be closer to reality. Although the reactive transport calculations presented below are based on the model we developed here, a combination of several thermodynamic models could be used through suitable choices of the kinetics of C-S-H reactions. However, in this case clearly the lowest solubility trend would eventually crystallize at the expense of the higher solubility trends.

Implementation of C-S-H Model in Reactive Transport Calculations

Most reactive transport calculations do not include the effect of solid solution behavior on the evolution of the system. In part, this is because these codes are designed to work with pure

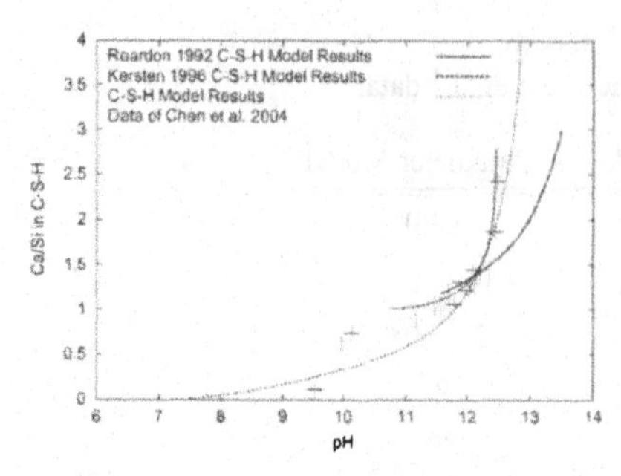

Figure 3: Measured and calculated variation of the Ca/Si ratio in C-S-H as a function of the aqueous solution pH.

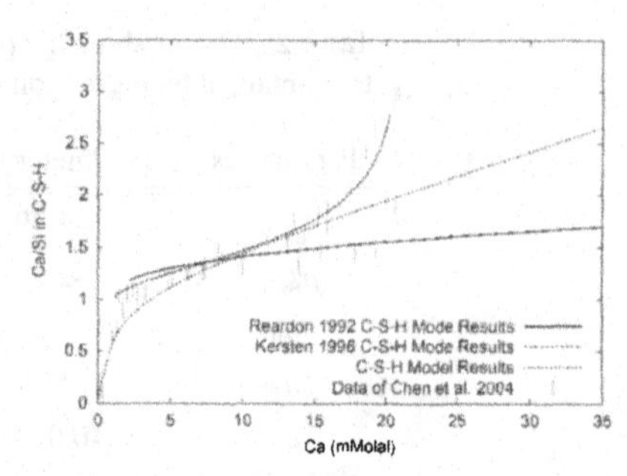

Figure 4: Measured and calculated variation of the Ca/Si ratio in C-S-H as a function of the aqueous solution Ca concentration.

phases and solid solutions would appear to require the use of additional variables and subroutines to calculate solid solution-aqueous solution equilibria. Recently, Lichtner and Carey [13] developed the theoretical and practical basis for incorporating solid solutions in reactive transport codes by discretizing the solid solution into a set of pure phases. They demonstrated that by implementing the discrete representation in a kinetically controlled reactive transport calculation the resulting system evolves toward the correct composition-dependent equilibrium state.

The discrete representation in combination with a kinetic formulation allows a flexible approach to modeling dissolution and precipitation behavior. In general, the discrete phases are averaged to represent the predicted composition of the solid solution. However, it is also possible to distinguish between primary and secondary phases. These features will be more obvious in the example calculations that follow.

For C-S-H, the solid solution is represented according to Equation 3 and the endmembers $Ca(OH)_2$ and SiO_2. In our calculations, we have used 100 discrete compositions so that the solid solution is represented by

$$\mathrm{Ca(OH)_2}, [\mathrm{Ca(OH)_2}]_{0.01} \cdot [\mathrm{SiO_2}]_{0.99}, \quad \ldots \quad [\mathrm{Ca(OH)_2}]_{0.99} \cdot [\mathrm{SiO_2}]_{0.01}, \mathrm{SiO_2}. \tag{10}$$

The solubility constant for each of these phases was calculated using [13]

$$K_{ss}(x) = \left(\frac{K_1}{\lambda_1(x)x}\right)^x \left(\frac{K_2}{\lambda_2(x)(1-x)}\right)^{1-x}. \tag{11}$$

Table 3 provides a subset of calculated solubility constants and activity coefficients. The discretization and solubility calculation is perfectly general and although we use the C-S-H formulation developed in Equation 5, other formulations can be readily discretized and, given a method of calculating the solubility constant as a function of composition (e.g., Equation 11), can be incorporated into a reactive transport calculation as described below.

Table 3: Calculated thermodynamic values for C-S-H solid solutions.

X_{SiO_2}	Ca/Si	$\log K_{sp}$	λ_{SiO_2}	$\lambda_{Ca(OH)_2}$
0.00	INF	-3.4900	0.0000	0.0000
0.05	19.0000	-4.1285	-26.9812	-0.0762
0.10	9.0000	-4.6697	-24.1696	-0.3041
0.15	5.6667	-5.1324	-21.5177	-0.6829
0.20	4.0000	-5.5208	-19.0244	-1.2117
0.25	3.0000	-5.8367	-16.6889	-1.8897
0.30	2.3333	-6.0815	-14.5104	-2.7160
0.35	1.8571	-6.2560	-12.4878	-3.6898
0.40	1.5000	-6.3607	-10.6204	-4.8102
0.45	1.2222	-6.3960	-8.9073	-6.0764
0.50	1.0000	-6.3624	-7.3475	-7.4875
0.55	0.8182	-6.2600	-5.9403	-9.0428
0.60	0.6667	-6.0890	-4.6847	-10.7414
0.65	0.5385	-5.8494	-3.5800	-12.5824
0.70	0.4286	-5.5411	-2.6253	-14.5652
0.75	0.3333	-5.1639	-1.8197	-16.6889
0.80	0.2500	-4.7174	-1.1625	-18.9527
0.85	0.1765	-4.2007	-0.6527	-21.3558
0.90	0.1111	-3.6122	-0.2895	-23.8974
0.95	0.0526	-2.9481	-0.0722	-26.5768
1.00	0.0000	-2.1900	0.0000	0.0000

MODELING CEMENT DEGRADATION

This work describes the implementation of a continuum model to describe the degradation of cement. A continuum model representing transport of reacting solutes in porous and fractured materials is based on a macro-scale representation of fluid flow, transport and reaction [14]. In this approach, micro-scale properties are represented through macro-scale properties such as porosity, permeability, tortuosity, and effective mineral surface area. Rates of heterogeneous chemical reactions taking place at solid-fluid interfaces are described by upscaled homogeneous rates averaged over a control volume. Continuum-scale approaches ignore small-scale heterogeneity and are insensitive to micro-scale features such as mineral texture. Nevertheless these models have been successful in simulating reactive transport in porous media (e.g. [14,15,16]).

Difficult challenges face any attempt to model quantitatively the degradation of cement. This is primarily due to the change in solid structure with time resulting in changes in material properties including porosity, permeability, tortuosity, and solid composition of the cement. To account for this evolution in pore geometry at the continuum scale a very simplistic approach is employed in which phenomenological constitutive relations are introduced that represent changes in material properties such as permeability and tortuosity as functions of porosity. The change in porosity is generally computed by adding up the volumes of the reacting solids within each control volume. However, this approach leads to additional difficulties in that, depending on the initial porosity and volume change of reaction, it possible for the porosity to become negative invalidating the calculation. When this occurs, it is an indication that other processes such as mechanical deformation must be taken into account. However, incorporation of mechanical effects is beyond the scope of this work. As a result of these limitations, continuum models are perhaps better suited to a more qualitative approach that can provide insight into the physical and chemical processes taking place.

Reactive Transport Model FLOTRAN

In this work the computer code FLOTRAN[17], a two-phase, nonisothermal, multicomponent, reactive flow and transport model, is applied to the problem of cement degradation. FLOTRAN is capable of describing two-phase flow and transport of water and gas phases coupled to multicomponent solute transport and reaction with aqueous and gaseous species and solids. The model includes both Pitzer and Debye-Hückel activity coefficient algorithms[15,18,19], a dual continuum implementation[20], colloid-facilitated transport, and other features.

The reactive transport model FLOTRAN consists of two distinct modules, FLOW90 and TRANS90, for flow and transport, respectively. These modules may be coupled sequentially to one another or run in stand-alone mode. The flow equations determine the temperature, pressure, saturation state, and velocity of each fluid phase. After completion of a flow time step, these values are passed to the reactive transport equations where a transport step is carried out taking into account multicomponent chemical reactions with aqueous and gaseous species, and solids. Changes in porosity due to solid reactions can then be coupled back to the flow equations providing for changes in permeability, for example. The reactive transport equations account for such processes as aqueous speciation, reactions with minerals, cation exchange and surface complexation, and colloid-facilitated transport. In this work only the reactive transport module TRANS90 is described in some detail.

Chemical Processes

Chemical reactions included in FLOTRAN consist of homogeneous aqueous reactions, and heterogeneous mineral precipitation and dissolution for both pure phases and solid solutions, ion exchange, surface complexation, and gaseous reactions. Homogeneous reactions taking place within the aqueous phase and heterogeneous aqueous-gaseous reactions can be written in the general form

$$\sum_{j=1}^{N_c} \nu_{ji}^{aq} \mathcal{A}_j^{aq} \rightleftharpoons \mathcal{A}_i^{aq}, \tag{12}$$

and

$$\sum_{j=1}^{N_c} \nu_{ji}^g A_j^{aq} \rightleftharpoons A_i^g, \tag{13}$$

for the ith complex or gaseous species A_i^α, $(\alpha = aq, g)$, where species A_j^{aq} refer to primary species assumed to be N_c in number, and ν_{ji}^α denotes the stoichiometric reaction matrix. In what follows, subscripts i and j are reserved for primary and secondary species, respectively, and subscripts k and l are reserved for either primary or secondary species. Primary species, selected from the set of aqueous species, serve as independent basis species to write the reactions. Their choice is arbitrary, and primary and aqueous secondary species may be interchanged freely so long as the resulting reactions (12) remain linearly independent[14]. Mineral reactions involving pure phases have the similar form

$$\sum_{j=1}^{N_c} \nu_{jm} A_j^{aq} \rightleftharpoons \mathcal{M}_m, \tag{14}$$

with stoichiometric reaction matrix ν_{jm} associated with mineral $\mathcal{M}_m$. For solid solutions, the discrete kinetic model recently introduced by Lichtner and Carey[13] is employed. In this approach the solid solution $\mathcal{M}_{x_{mk}}$ characterized by the composition $x_m = \{x_1^m, \ldots, x_{N_m}^m\}$ with mole fractions x_i^m, is discretized into a sequence of stoichiometric solids with compositions labeled $x_{mk} = \{x_{1k_1}^m, \ldots, x_{N_m k_{Z_m}}^m\}$, where the subscripts $k_1, \ldots, k_{Z_m}$ refer to a particular discretization with Z_m different stoichiometric solids represented by the simultaneous set of reactions

$$\sum_{ji} x_{ik}^m \nu_{ji}^m A_j^{aq} \rightleftharpoons \mathcal{M}_{x_{mk}}. \tag{15}$$

Reaction rates involving aqueous complexes can generally be assumed to be sufficiently fast compared to advective and diffusive mass transport that local equilibrium is an adequate approximation. Concentrations of secondary species in local chemical equilibrium are obtained as nonlinear functions of the concentrations of primary species C_j^{aq} from the law of mass action as

$$C_i^\alpha = (\lambda_i^\alpha)^{-1} K_i^\alpha \prod_{j=1}^{N_c} (\lambda_j^{aq} C_j^{aq})^{\nu_{ji}^\alpha}, \tag{16}$$

with equilibrium constant K_i^α associated with the ith secondary species and αth phase.

Mineral reactions involve mass transfer between the aqueous and solid phases. For pure mineral phases and discretized solid solutions their rates are described through a kinetic rate law of the form

$$I_{x_{mk}} = -k_{x_{mk}} a_{x_{mk}} \mathcal{P}_{x_{mk}} \left[1 - K_{x_{mk}} Q_{x_{mk}}\right] \zeta_{x_{mk}}(\phi_{x_{mk}}, Q_{x_{mk}}), \tag{17}$$

provided by transition state theory. For a pure mineral phase, $x_{mk} = 1$. In this expression, $k_{x_{mk}}$ denotes the kinetic rate constant in general a function of the solid solution composition, $a_{x_{mk}}$ refers to the mineral surface area per bulk volume of porous medium, $K_{x_{mk}}$ denotes to the equilibrium constant for reaction (14) for pure mineral phases and reaction (15) for solid solutions, and $Q_{x_{mk}}$ denotes the ion activity product defined by

$$Q_{x_{mk}} = \prod_{j=1}^{N_c} (\lambda_j^{aq} C_j^{aq})^{\nu_{jm}^{aq}}. \tag{18}$$

The prefactor $\mathcal{P}_m$ accounts for pH and other dependencies of the reaction rate with the form

$$\mathcal{P}_{x_{mk}} = \prod_{k=1}^{N} a_k^{\sigma_{mk}}, \tag{19}$$

in which the activity of the kth solute species a_k is raised to an integral or nonintegral power σ_{mk}. The factor $\zeta_{x_{mk}}$ takes on the values zero or one to ensure that if a mineral is not present at some particular point in space, it cannot dissolve:

$$\zeta_{x_{mk}} = \begin{cases} 1, & \phi_{x_{mk}} > 0 \text{ or } K_{x_{mk}} Q_{x_{mk}} > 1, \\ 0, & \phi_{x_{mk}} = 0 \text{ and } K_{x_{mk}} Q_{x_{mk}} < 1 \end{cases}. \tag{20}$$

The quantity in square brackets on the right hand side of Eqn. (17) is referred to the affinity factor, and vanishes at equilibrium. The sign of the affinity factor determines whether precipitation or dissolution occurs, with a positive rate designating precipitation corresponding to reaction (14) proceeding from left to right.

Mass and Energy Conservation Equations

In terms of the general forms of homogeneous and heterogeneous reactions described above, the computer code FLOTRAN[17] solves the following transient, two-phase, mass conservation equations for a set of primary solute species, and minerals

$$\frac{\partial}{\partial t}\left[\phi\left(s_{aq}\Psi_j^{aq} + s_g\Psi_j^g\right)\right] + \boldsymbol{\nabla}\cdot\left[\boldsymbol{\Omega}_j^{aq} + \boldsymbol{\Omega}_j^g\right] = S_j - \sum_m \nu_{jm}\omega_{x_{mk}} I_{x_{mk}}, \tag{21}$$

and

$$\frac{\partial \phi_{x_{mk}}}{\partial t} = \overline{V}_{x_{mk}}\omega_{x_{mk}} I_{x_{mk}}. \tag{22}$$

In these equations s_α refers to liquid water saturation of phase $\alpha = aq,\, g$ with

$$s_{aq} + s_g = 1. \tag{23}$$

The quantities Ψ_j^α, $\boldsymbol{\Omega}_j^\alpha$ denote the total concentration and flux, respectively, in liquid and gas phases defined by

$$\Psi_j^\alpha = \delta_{aq,\alpha} C_j^{aq} + \sum_i \nu_{ji} C_i^\alpha, \tag{24}$$

and

$$\boldsymbol{\Omega}_j^\alpha = \delta_{aq,\alpha} \boldsymbol{J}_j^{aq} + \sum_i \nu_{ji}^\alpha \boldsymbol{J}_i^\alpha, \tag{25}$$

respectively, where the sum over i runs over secondary species, with the solute flux $\boldsymbol{J}_k^\alpha$ for individual aqueous and gaseous species defined by

$$\boldsymbol{J}_k^\alpha = \boldsymbol{q}_\alpha C_k^\alpha - \phi s_\alpha \tau_\alpha D_\alpha \boldsymbol{\nabla} C_k^\alpha. \tag{26}$$

This formulation of aqueous diffusion based on a single diffusion coefficient is highly simplistic for high ionic strength electrolyte solutions[21,22]. To ensure convergence of the solid solution discretization, the kinetic rate terms are multiplied by the weight factor $\omega_{x_{mk}}$, $\sum_k \omega_{x_{mk}} = 1$, equal to the number of dissolving or precipitating stoichiometric solids within a control volume[13]. For pure phase minerals $\omega_{x_{mk}} = 1$. The effective gaseous diffusion coefficient has the form

$$D_g = \lambda_g D_g^0 \frac{p_{\rm ref}}{p_g} \left(\frac{T + T_{\rm ref}}{T_{\rm ref}}\right)^{1.8}, \tag{27}$$

where D_g^0 refers to the diffusion coefficient in a pure gas phase, λ_g denotes an enhancement factor, and $p_{\rm ref}$ and $T_{\rm ref}$ refer to reference pressure and temperature. The quantity S_j represents a source/sink term.

The mineral mass transfer equation, Eqn.(22) is expressed in terms of the mineral volume fraction $\phi_{x_{mk}}$. The quantity $\overline{V}_{x_{mk}}$ denotes the mineral molar volume. The total porosity is related to the mineral volume fractions by the expression

$$\phi = 1 - \sum_{mk} \phi_{x_{mk}}. \tag{28}$$

It should be kept in mind that this formulation does not distinguish between connected and disconnected pore space, which is certainly an important distinction in cement and concrete alteration.

Species-dependent diffusion coefficients. The mass transport equations do not conserve charge with the flux as written in the form of Equation (26) if the diffusion coefficients are different for differently charged ions. This is because different ions can diffuse at different rates depending on their diffusion coefficients. For example, the hydrated hydrogen ion has a diffusion coefficient which is an order of magnitude larger compared to the typical value of 10^{-5} cm^2/s[23]. The hydroxyl species OH^- has a diffusion coefficient roughly five times larger than the typical value. A list of diffusion coefficients for some of the common cations, anions and neutral species is given in Table 4.

In order to obtain transport equations which conserve charge in this case, an additional term must be added to the solute flux providing for electrochemical migration[23,24]. Electrochemical migration is caused by the presence of an electric field which is produced locally by the different rates of diffusion of differently charged ions. The solute flux $\boldsymbol{J}_i$, in general, consists of contributions from advective, diffusive and dispersive transport in addition to an electrochemical migration term. Ignoring dispersion, the flux has the form[23,24].

$$\boldsymbol{J}_i = -\tau\phi L_i \boldsymbol{\nabla}\mu_i + \mathbf{q} C_i, \tag{29}$$

where the chemical potential μ_i has the form

$$\mu_i = \mu_i^\circ + RT \ln \lambda_i C_i + z_i \mathcal{F}\psi, \tag{30}$$

where μ_i° refers to the standard state potential, the quantity ψ represents the electrical potential, τ refers to the tortuosity of the porous medium, and q denotes the Darcy fluid velocity. The superscript aq is dropped for convenience since all species are considered to be in the aqueous phases in what follows. The coefficient L_i is related to the usual diffusion coefficient D_i by the expression

$$L_i = \frac{D_i C_i}{RT}. \tag{31}$$

Table 4: List of species-dependent diffusion coefficients for a selected set of species at 25°C. Values are multiplied by 10^5.

Cations	D_i [cm^2/s]	Anions	D_i [cm^2/s]	Neutral	D_i [cm^2/s]
Cs^+	2.1	F^-	2.1	$O_{2(aq)}$	2.09
Cu^+	1.2	Cl^-	2.032	$H_{2(aq)}$	2.09
H^+	9.312	HCO_3^-	0.6		
K^+	1.957	HSO_4^-	1.33		
Na^+	1.334	I^-	2.044		
NH_4^+	1.954	OH^-	5.26		
Cu^{2+}	0.8	SO_4^{2-}	1.065		
Ca^{2+}	0.792				
Fe^{2+}	0.8				
Mg^{2+}	0.706				
Ni^{2+}	0.8				
Sr^{2+}	0.9				
Zn^{2+}	0.8				

This expression should be compared to Eqn. (12) in Samson et al.[25] in which the concentration factor C_i was inadvertently omitted. Note that a more general expression would also include off diagonal terms in Eqn. (29). Substituting Eqn. (29) in Eqn. (26) results in the following form of the solute flux

$$\boldsymbol{J}_i = -\tau\phi z_i \frac{D_i C_i}{RT}\mathcal{F}\boldsymbol{\nabla}\psi - \tau\phi D_i\left(\boldsymbol{\nabla}C_i + C_i\boldsymbol{\nabla}\ln\lambda_i\right) + \mathbf{q}C_i, \tag{32}$$

where the first term refers to electrochemical migration, the second term to aqueous diffusion, and the last term to advective transport. Here z_i, λ_i and D_i denote the charge, activity coefficient and diffusivity of the ith species, respectively. The electrochemical migration term is proportional to the ion charge, the electric field ($\boldsymbol{E} = -\boldsymbol{\nabla}\psi$), and the mobility D_iC_i/RT. The potential ψ is determined in such a manner as to ensure charge conservation as discussed below[24].

With species-dependent diffusion coefficients, and taking into account corrections due to activity coefficients, the expression for the generalized flux, Equation (25), can be written in the form[26]

$$\boldsymbol{\Omega}_j = -\mathcal{F}\tau\phi\Psi_j^{\epsilon}\frac{\boldsymbol{\nabla}\psi}{RT} - \tau\phi\left(\boldsymbol{\Gamma}_j^D + \boldsymbol{\Gamma}_j^{\lambda}\right) + \mathbf{q}\boldsymbol{\Psi}_j, \tag{33}$$

obtained by inserting the expression for the flux defined in Equation (32) into equation Equation (25). The quantities Ψ_j^{ϵ} and $\boldsymbol{\Gamma}_j^D$ are defined by the expressions

$$\Psi_j^{\epsilon} = z_j D_j C_j + \sum_i \nu_{ji} z_i D_i C_i, \tag{34}$$

and

$$\boldsymbol{\Gamma}_j^D = D_j\boldsymbol{\nabla}C_j + \sum_i \nu_{ji} D_i \boldsymbol{\nabla}C_i. \tag{35}$$

Activity coefficient corrections are accounted for by the term containing $\mathbf{\Gamma}_j^\lambda$ defined by

$$\mathbf{\Gamma}_j^\lambda = D_j C_j \mathbf{\nabla} \ln \lambda_j + \sum_i \nu_{ji} D_i C_i \mathbf{\nabla} \ln \lambda_i. \tag{36}$$

Electroneutrality. One of the fundamental properties of an aqueous solution is that it is electrically neutral on a macroscopic scale. Therefore, for the mass transport equations to properly represent such systems, they must conserve charge. The charge balance equation has the general form

$$\frac{\partial}{\partial t}(\phi \rho_Q) + \mathbf{\nabla} \cdot \boldsymbol{i} = 0. \tag{37}$$

obtained by multiplying Eqn. (21) by z_j and summing over all primary species, where the total charge and current densities, ρ_Q, $\boldsymbol{i}$, respectively, are defined by

$$\rho_Q = \mathcal{F} \sum_{i=1}^{N} z_i C_i = \mathcal{F} \sum_{j=1}^{N_c} z_j \Psi_j, \tag{38}$$

and

$$\boldsymbol{i} = \mathcal{F} \sum_{i=1}^{N} z_i \boldsymbol{J}_i = \mathcal{F} \sum_{j=1}^{N_c} z_j \boldsymbol{\Omega}_j. \tag{39}$$

The Faraday constant $\mathcal{F}$ has the value $\mathcal{F}$ = 96487 coulombs/mole, giving ρ_Q units of coulombs per unit volume, and the current density units of coulombs per unit area per second, or amperes per unit area (1 ampere = 1 coulomb/second). The right hand side of Eqn. (37) vanishes because chemical reactions conserves charge and hence

$$\sum_{jm} z_j \nu_{jm} I_m = \sum_m z_m I_m = 0, \tag{40}$$

since solids have no bulk charge: $z_m = 0$.

For the case of species-dependent diffusion coefficients, demanding that the solution current density $\boldsymbol{i}$ defined by Equation (39) vanish identically, requires that the gradient in the electric potential $\mathbf{\nabla}\psi$ be equal to

$$-\mathbf{\nabla}\psi = \frac{\tau \phi \mathcal{F}}{\kappa} \sum_j z_j \left(\mathbf{\Gamma}_j^D + \mathbf{\Gamma}_j^\lambda\right), \tag{41}$$

obtained by writing out Equation (39) using Equation (33), and solving for $\mathbf{\nabla}\psi$. The inverse of the quantity κ, defined by the expression

$$\kappa = \frac{\tau \phi \mathcal{F}^2}{RT} \sum z_j \Psi_j^\epsilon, \tag{42}$$

is referred to as the generalized Debye length. For concentrated solutions the Debye length κ^{-1} becomes small and the effects of the potential negligible. Using Equation (41) to eliminate the potential ψ, the generalized flux may be written as

$$\boldsymbol{\Omega}_j = -\tau \phi \sum_l \beta_{jl} \left(\mathbf{\Gamma}_l^D + \mathbf{\Gamma}_l^\lambda\right) + \boldsymbol{q} \Psi_j. \tag{43}$$

The coefficient matrix β_{jl} is a projection operator ($\beta^2 = \beta$) defined by

$$\beta_{jl} = \delta_{jl} - z_l \omega_j, \tag{44}$$

with

$$\omega_j = \frac{\Psi_j^\epsilon}{\sum_l z_l \Psi_l^\epsilon}. \tag{45}$$

The matrix β_{jl} has the property that

$$\sum_j z_j \beta_{jl} = 0, \tag{46}$$

and thus

$$\boldsymbol{i} = \sum z_j \boldsymbol{\Omega}_j = 0. \tag{47}$$

Thus in contrast to the expression for the flux with species-independent diffusion coefficients, coupling terms now occur between the concentration gradients of the different primary species. These terms are required to maintain electrical balance within the aqueous solution. Note that in this formulation there is no need to solve Poisson's equation to obtain the potential ψ. In fact the actual value of ψ is not needed, only its gradient appears in the Nernst-Planck equation which, as demonstrated above, can be obtained directly from the condition of electroneutrality.

Making use of the above choice of the electrical field so that the current density vanishes identically, it follows that the charge conservation equation for constant porosity reduces to

$$\frac{\partial}{\partial t}(\phi \rho_Q) = \phi \frac{\partial}{\partial t}(\rho_Q) = 0. \tag{48}$$

As a consequence for electrically neutral initial and boundary conditions, $\rho_Q = 0$ for all time, and charge is conserved.

Debye-Hückel and Pitzer Activity Coeffi cient Algorithm

For relatively low ionic strength fluids, defined as $I \lesssim 0.1\ m$, where I denotes the ionic strength of the fluid given by

$$I = 1/2 \sum_k z_k^2 m_k, \tag{49}$$

with the molality of the kth species denoted by m_k with valence z_k, the extended Debye-Hückel activity coefficient λ_k defined by

$$\log \lambda_k(I) = -\frac{z_k^2 A(T)\sqrt{I}}{1 + \mathring{a}_k\ B(T)\sqrt{I}} + \dot{b}(T) I, \tag{50}$$

is adequate for computing activity coefficient corrections. In this expression $A(T), B(T), \dot{b}(T)$ refer to temperature dependent coefficients, and $\mathring{a}_k$ denotes the Debye radius. The activity of the solvent, in this case water, is unity for an ideal dilute solution. In the Debye-Hückel algorithm the activity coefficient is a function of ionic strength I, and is the same for species with identical valencies and Debye radii $\mathring{a}_k$.

For high ionic strength fluids, the Debye-Hückel model is not adequate. For such fluids an approach such as the Pitzer model may be needed[27]. In this approach the activity coefficients are expressed in a virial expansion of the form

$$\ln \lambda_k = \ln \lambda_k^0 + \sum_{k'} \mathcal{B}_{kk'}(I)\, m_{k'} + \sum_{k'}\sum_{k''} \mathcal{C}_{kk'k''}\, m_{k'}\, m_{k''} + \cdots, \tag{51}$$

where λ_k refers to the individual ion activity coefficient[18,27], and λ_k^0 denotes a modified form of the Debye-Hückel activity coefficient. The expansion coefficients, $\mathcal{B}_{kk'}$ and $\mathcal{C}_{kk'k''}$, must be determined through fits to experimental measurements over a range of pressure and temperature conditions. It is generally not possible to extrapolate results beyond the region used for fitting the data. The activity of water is calculated from the relation

$$\ln a_w = -RT\, W_w\, \emptyset \sum_{k \neq w} m_k, \tag{52}$$

where R denotes the gas constant, T is temperature, W_w refers to the formula weight of water, and $\emptyset$ denotes the osmotic coefficient of water (see[27]).

The Pitzer activity coefficients account for electrostatic interaction and ion hydration effects. In addition, they account for formation of complexes and ion pairing with the exception of strong complexes such as $CaSO_4(aq)$, for example. Because activity coefficients in the Pitzer formulation account for aqueous complexing and ion pairing, the greater these effects the smaller the activity coefficients, thereby reducing the species activity. In this sense the Pitzer approach differs conceptually from the Debye-Hückel formulation in which only electrostatic and hydration effects are included. In the Debye-Hückel formulation, complexing and ion pairing must be explicitly taken into account. When using a Debye-Hückel activity coefficient algorithm in speciation calculations any of the relevant chemical species that are found in a thermodynamic database are generally employed. Often many of these species are unimportant, but some are essential. The Pitzer model, in contrast, uses a restricted set of aqueous species compared to the Debye-Hückel model. To determine which additional species are necessary to include in any particular chemical system, the user must refer to the original data on which the expansion coefficients in the Pitzer model are based. For example, the species $NaCl_{(aq)}$ is not considered a valid species in the Pitzer model at moderate temperatures, but is usually included in a Debye-Hückel model although at low temperatures it probably has little effect within the range of validity of the Debye-Hückel algorithm. On the other hand, strong complexes such as $NaNO_{3(aq)}$ and $NaNO_{2(aq)}$ are required in both models for an accurate representation of the fluid composition. In addition, species which undergo protonation and deprotonation, such as $H^+ - OH^- - H_2O$, $CO_{2(aq)} - CO_3^{2-} - HCO_3^-$, $Al^{3+} - Al(OH)_4^-$, $Fe^{3+} - Fe(OH)_4^-$, and $SiO_{2(aq)} - H_3SiO_4^- - H_2SiO_4^{2-}$ generally must be explicitly taken into account.

Numerical Solution

The solute transport equations are solved numerically using a fully implicit backward Euler algorithm. For 1D systems a tridiagonal solver is used. For 2D and 3D systems an iterative solver is employed using GMRES. The mineral mass transfer equations are solved using an explicit finite difference procedure using the kinetic reaction rates for minerals obtained from the solution to the transport equations. This latter simplification is possible because of the slow rates of reaction associated with solids resulting in the approximate decoupling of the mineral and solute conservation equations over a single time step.

To derive the finite volume discretized equations consider a partial differential equation of the general form

$$\frac{\partial A}{\partial t} + \boldsymbol{\nabla} \cdot \boldsymbol{F} = \mathcal{S}, \tag{53}$$

with accumulation term A, source/sink term $\mathcal{S}$, and flux term $\boldsymbol{F}$ of the form

$$\boldsymbol{F} = q\rho X - \phi D\rho \boldsymbol{\nabla} X. \tag{54}$$

Partitioning the computational domain into a set of finite volumes V_n and integrating the partial differential equations over each volume yields a discretized form of the mass conservation equations. The following results are obtained for the accumulation, source, and flux terms:

$$\int_{V_n} \frac{\partial}{\partial t} A \, dV \simeq \frac{A_n^{t+\Delta t} - A_n^t}{\Delta t} V_n, \tag{55}$$

for the accumulation term,

$$\int_{V_n} \mathcal{S} \, dV \simeq \mathcal{S}_n V_n, \tag{56}$$

for the source term, and

$$\int_{V_n} \boldsymbol{\nabla} \cdot \boldsymbol{F} \, dV = \int_{\partial V_n} \boldsymbol{F} \cdot \boldsymbol{dS} = \sum_{n'} F_{nn'} A_{nn'}, \tag{57}$$

for the flux term, where ∂V_n denotes the surface of V_n, and the sum is over the neighboring volumes connected to V_n. The flux $F_{nn'}$ across the $n-n'$ interface connecting volumes V_n and $V_{n'}$ is given by

$$F_{nn'} = (q\rho)_{nn'} X_{nn'} - (\phi D\rho)_{nn'} \frac{X_n - X_{n'}}{d_n + d_{n'}}, \tag{58}$$

where the subscript nn' indicates that the quantity is evaluated at the interface connecting nodes n and n', the quantities d_n, $d_{n'}$ denote the distances from the centers of the control volumes V_n, $V_{n'}$ to the their common interface with interfacial area $A_{nn'}$. The quantity $X_{nn'}$ is usually computed using upwinding to avoid instabilities in the finite volume. Combining these results gives the residual equation for the discretized form of the partial differential equations

$$R_n^k = (A_n^k - A_n^{k-1}) \frac{V_n}{\Delta t} + \sum_{n'} F_{nn'}^k A_{nn'} - \mathcal{S}_n^k V_n, \tag{59}$$

for the kth time step, where, in general, R_n^k is a nonlinear function of the independent field variables. These equations may be solved using a Newton-Raphson iteration technique in which the discretized equations are first linearized resulting in the Newton-Raphson equations

$$\sum_{n'} J_{nn'}^i \, \delta x_{n'}^{i+1} = -R_n^i, \tag{60}$$

for the i iterate, with the Jacobian matrix $J_{nn'}^i$ defined by

$$J_{nn'}^i = \frac{\partial R_n^i}{\partial x_{n'}^i}. \tag{61}$$

To account for large spatial changes in concentration of a given species, e.g. changes in speciation due to large changes in pH, it has been found useful to solve for the logarithm of the concentration. Thus $x_n = \ln C_{jn}$, and

$$\delta x_n^{i+1} = \ln C_{jn}^{i+1} - \ln C_{jn}^{i}. \tag{62}$$

After each iteration the solution is updated through the equation

$$\ln C_{jn}^{i+1} = \ln C_{jn}^{i} + \delta x_n^{i+1}, \tag{63}$$

or

$$C_{jn}^{i+1} = C_{jn}^{i} e^{\delta x_n^{i+1}}. \tag{64}$$

After the transport equations have been solved over a time step, mineral concentrations are obtained directly from the mineral mass transfer equations at each node n according to the relation:

$$\phi_{mn}(\boldsymbol{r}, t + \Delta t) = \phi_{mn}(\boldsymbol{r}, t) + \Delta t \overline{V}_m I_{mn}(\boldsymbol{r}, t), \tag{65}$$

where the mineral reaction rate $I_{mn}(\boldsymbol{r}, t)$ is taken from the previous time step. The time-step size Δt depends on how close the solution to the aqueous and gaseous species transport equations is to a stationary state[28]. For the transient case, the same time step Δt is used as in the transient transport equations. When the system has reached a stationary state, however, a much larger time step can be taken without fear of violating the stability conditions. In this case, the time-step size is only restricted by the maximum absolute change allowed in the mineral volume fraction. The allowable time-step size is also controlled by limiting the negative most value of the mineral volume fraction to an acceptable value which may be taken as zero. Note, however, that at a reaction front where the mineral volume fraction vanishes, according to Eq. (65) in order for the front to move, the volume fraction must become negative over some region surrounding the front. This is, of course, nonphysical and the volume fraction must be set back to zero to eliminate the negative values, if they occur. In the transient regime, the change in mineral abundances and, hence, porosity and permeability, is generally much slower than the change in aqueous and gaseous compositions as well as temperature, pressure, and saturation.

APPLICATIONS

In the following, we investigate the predictive capabilities of the FLOTRAN reactive transport model for simulating cement degradation by carbonation and sulfate attack. In the case of carbonation, we make use of a set of observational data collected on oilwell cements and examine the sensitivity of the model results to assumed model parameters. For sulfate attack, our analysis is limited to a brief evaluation of idealized sulfate degradation processes.

Carbonation of Oilwell Cement

There is renewed interest in the longterm stability of oilwell cements in a CO_2-rich environment because of the possibility of reducing greenhouse gas emissions by storing CO_2 in the subsurface in a process known as geologic sequestration[29]. However, storage of CO_2 is expected to produce CO_2-saturated brine that could attack the cement that provides a primary seal against fluids escaping the geologic reservoir. There is a critical need to provide model predictions of the performance of the cement wellbore seal for the long periods (100's of years) required in the geologic sequestration of CO_2.

In a recent field study conducted in collaboration with Kinder Morgan CO_2 Company, L.P.[1], we recovered samples from an enhanced-oil recovery field in West Texas with 30 years of CO_2 exposure. The cement core was collected 3 m above the CO_2-bearing limestone reservoir and illustrates some of the potential carbonation reactions that can occur in the wellbore environment (Figure 5). We observed a zone of mildly carbonated, gray-colored cement (that retains portlandite, hydrogarnet, and Friedel's salt) adjacent to the casing, and a zone of intensely carbonated, orange-colored cement (containing only calcite, aragonite, and vaterite as crystalline phases) adjacent to the shale caprock[30]. Our analysis indicates that CO_2-saturated fluids migrated from the reservoir along the cement-shale interface, possibly because this interface had relatively high porosity/permeability. CO_2 from the interface diffused into the cement, producing the orange carbonated zone.

Table 5: Quantitative X-ray diffraction data (in weight percent) for cement samples recovered from well 49-6 at the SACROC unit, West Texas (Figure 5). All crystalline phases normalized to 100%. Where determined, the quantity of amorphous material was obtained using an internal reference standard.

	Sample ID				
Depth (ft)	6543	6545	6549	6549	6549
Phase	Gray Cement	Gray Cement	Gray Cement	Orange Cement	Orange Cement
Portlandite	33.0	45.3	38.8	—	—
Hydrogarnet	25.7	33.3	36.7	—	—
Friedel's Salt	3.9	4.0	4.3	—	—
Ettringite	3.2	—	—	—	—
Brucite	2.5	—	—	—	—
Brownmillerite	—	9.3	10.8	—	—
Ca_2SiO_4	—	3.3	—	—	—
Calcite	—	2.0	7.2	43.2	49.5
Aragonite	—	—	—	42.8	30.9
Vaterite	—	—	—	12.4	18.2
Halite	31.8	3.3	2.2	1.6	1.5
Amorphous	major	85.0	86.1	44.2	47.9

The carbonation at SACROC is associated with a relatively small enrichment of SiO_2 and

[1] We thank Scott Wehner and Mike Raines for collecting samples and for discussion and analysis of sample history.

Oxide	Gray Cement	Orange Cement
SiO_2	22.40	26.09
TiO_2	0.24	0.26
Al_2O_3	4.63	5.20
Fe_2O_3	2.52	2.74
MnO	0.11	0.10
MgO	3.46	1.36
CaO	62.94	58.90
Na_2O	2.24	4.29
K_2O	n.d	0.19
P_2O_5	0.10	0.10
LOI	18.67	35.26
Total	98.59	99.23

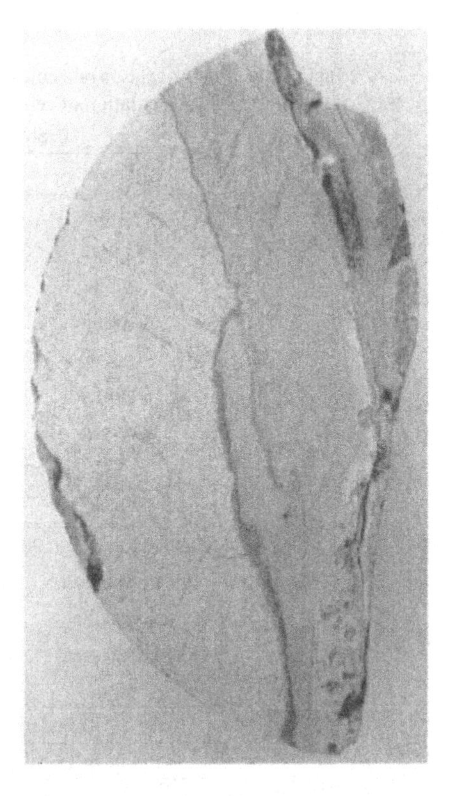

Figure 5: Polished slab of core from well 49-6 in the SACROC Unit, West Texas with a table of compositional data determined by x-ray fluorescence. The slab is oriented parallel to the wellbore (2.5 cm wide) with the casing on the left and the shale country rock on the right. Relatively unaltered gray cement with calcite-filled veins occurs adjacent to the casing and is separated from a heavily carbonated orange altered cement by a dark-gray translucent deposition zone.

depletion of CaO and a more significant enrichment in Na_2O and depletion in MgO (Figure 5). At the interface between the mildly carbonated gray cement and the intensely carbonated orange cement, a narrow, dark silicified zone appears to act as an impediment to further ingress of CO_2. At the cement-shale interface, the mineralogy is complex and includes veins of silica and carbonate in a matrix of intimately mixed carbonated cement and shale debris.

As an initial step in developing a predictive model of cement degradation in the wellbore environment, we have developed a simplified one-dimensional model of the carbonation process. Our conceptual model is that CO_2-saturated brine percolated up the cement-shale interface and in so-doing equilibrated with the shale mineralogy. As this fluid passes by the wellbore cement, CO_2-saturated brine diffuses into the cement matrix. In the model this process is represented with an initial state consisting of a column of shale and CO_2-saturated brine in contact with a fully cured, CO_2-free cement paste (Table 6). The model calculations were conducted at 25 °C (due to thermodynamic database limitations), although the SACROC reservoir temperature is near 50 °C. The brine in the shale was saturated with CO_2 at 179 bars, corresponding to the current field conditions. The brine chemistry for the shale was taken from a subset of the database developed by NETL[31] (Table 7). The initial cement fluid chemistry was governed by cement-phase equilibria. Zero gradient concentration boundary conditions were imposed at both ends of the column. The column was 0.25 m in length with cement occupying 0 – 0.05 m and shale with CO_2-saturated brine occupying 0.05 – 0.25 m.

Table 6: Minerals, reaction rate constants, and initial volume fractions used in the reactive transport calculation of cement carbonation in a wellbore environment.

Shale		
Phase	Volume Fraction	Rate[1]
Illite	0.198	1.E-30
Quartz	0.0659	1.E-20
Albite	0.01319	1.E-20
Kaolinite	0.00989	1.E-20
Calcite	0.00659	1.E-10
Dolomite	0.00659	1.E-12
Porosity	0.70	
Cement		
Phase	Volume Fraction	Rate[1]
C-S-H ($X_{SiO_2} = 0.36$)	0.379	5.E-11
Portlandite	0.154	1.E-10
Monosulfate	0.135	1.E-12
Hydrogarnet	0.033	1.E-12
Porosity	0.30	
Secondary Phases		
Phase	Volume Fraction	Rate[1]
Amorphous Silica	0	1.E-12
Gypsum	0	1.E-11
Gibbsite	0	1.E-10
Friedel's Salt	0	1.E-12
Ettringite	0	1.E-12
Brucite	0	1.E-12
Dawsonite	0	1.E-20
Magnesite	0	1.E-13

[1]mol/sec (The product of the intrinsic rate constant and surface area.)

The primary variables we investigated in the reactive transport simulation of this carbonation process included the porosity, the occurrence of solid solution in C-S-H, tortuosity, and mineral reaction kinetics. In addition, we examined the sensitivity of the results to the use of a species-dependent aqueous diffusion and the use of Pitzer activity coefficients. We conducted the reactive transport modeling with the aim of reproducing the following observational features from the SACROC sample (cf. Figure 5):

Table 7: Initial fluid compositions (in molality or as governed by mineral equilibrium) used in the reactive transport calculations of wellbore cement carbonation.

Shale	
Species	Concentration or Constraint
$Al(OH)_4^-$	2.8e-15
Ca^{2+}	Calcite
Mg^{2+}	Dolomite
Na^+	1.3089
K^+	0.1089
H^+	Charge Balance
HCO_3^-	179 bar CO_2 Pressure
$SiO_2(aq)$	Quartz
SO_4^{2-}	0.00624
Cl^-	1.661
Cement	
Species	Concentration or Constraint
$Al(OH)_4^-$	Hydrogarnet
Ca^{2+}	Portlandite
Mg^{2+}	Brucite
Na^+	1.e-12
K^+	1.e-12
H^+	Charge balance
HCO_3^-	1.e-12
$SiO_2(aq)$	CSH0.36
SO_4^{2-}	Monosulfate
Cl^-	1.e-12

- The width of the carbonation reaction zone was ≈ 0.5 cm and developed over a 30 year period
- The carbonation zone contains calcium carbonate and an amorphous alumino-silica residue
- The carbonation process resulted in only minor chemical changes, with a slight enrichment in SiO_2 and depletion in CaO

We consider initial porosity and tortuosity as shown in Table 8 with results given for the base case (Case 32) in Figures 6-9. After 30 years, the simulation produces a reaction zone of approximately 0.5 cm width that is marked by loss of all primary cement phases (C-S-H, monosulfate, portlandite, hydrogarnet) and the formation of secondary C-S-H, amorphous silica, calcite, gibbsite, and gypsum. In the broadest terms, the loss of primary cement phases results in the liberation of Ca, Si, and Al to precipitate as calcite, amorphous silica, and amorphous aluminum hydroxide (represented by gibbsite in the model).

Table 8: Conditions used in the carbonation simulations discussed in the text.

Case	Porosity Cement	Porosity Shale	Tortuosity	
32	0.30	0.70	0.0004	
40	0.30	0.30	0.0004	
42	0.30	0.09	0.0004	
34	0.30	0.70	0.004	
35	0.30	0.70	0.00004	
36	0.30	0.70	0.0004	No C-S-H Solid Solution
37	0.30	0.70	0.0004	No C-S-H Solid Solution or SiO_2(am)
38	0.30	0.70	0.0004	All kinetic rates $\div$ 100
39	0.30	0.70	0.0004	All kinetic rates $\times$ 100
44	0.30	0.70	0.0004	Species-dependent diffusion coefficients

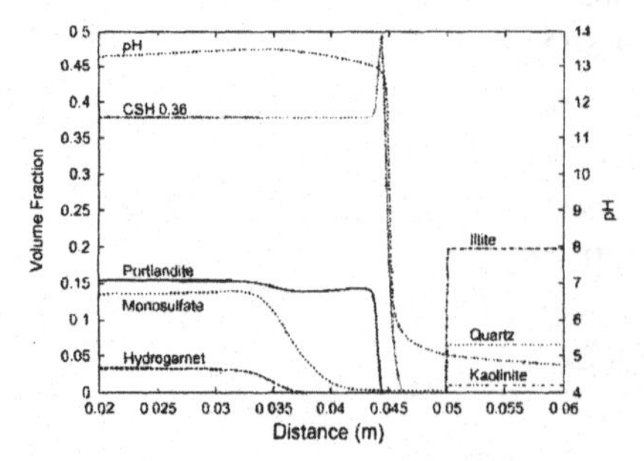

Figure 6: Calculated mineralogical and pH profile across the carbonation zone for the primary cement and country rock phases using case 32 at 30 years (Table 8). The initial cement-shale interface is at 0.05 m. None of the illustrated primary phases exist in the alteration zone between ≈ 0.046 - 0.05 m.

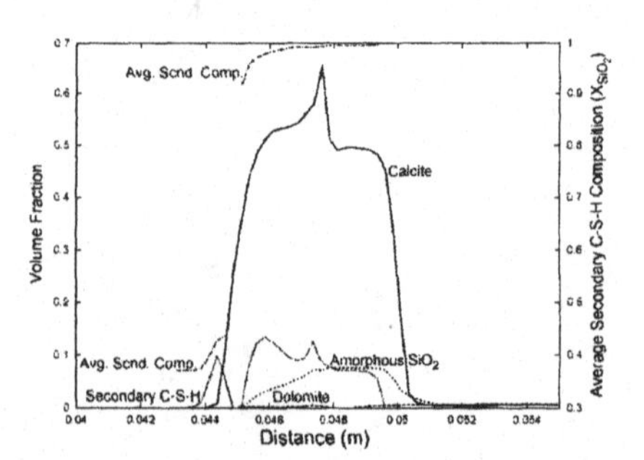

Figure 7: Calculated mineralogical profile across the carbonation zone for secondary carbonate and siliceous minerals using case 32 at 30 years (Table 8). The initial cement-shale interface is at 0.05 m. Note discontinuity in the average composition of the secondary C-S-H precipitate.

The model results reproduce many of the observations made on the recovered core shown in Figure 5 and Table 5: X-ray diffraction indicates that all of the observed cement phases (portlandite, hydrogarnet, Friedel's salt, and ettringite) are consumed by the carbonation process and replaced by a mixture of calcium carbonate polymorphs. However, because C-S-H is an amorphous, very fine-grained phase we have not been able to make comparisons of C-S-H distribution and composition between model predictions and observations. We believe that as in the model predictions, the C-S-H in the reaction zone is either highly silicified or completely replaced by amorphous silica and calcium carbonate.

The predicted compositional variation across the reaction zone (Figure 9) is in rough agreement with the x-ray fluorescence data of the bulk composition of the cement (Figure 5). The observed primary changes are only in H_2O and CO_2 with a small SiO_2 enrichment and CaO depletion in the carbonated zone. In contrast, the model predicts a small average depletion in SiO_2 and enrichment in CaO. The small-scale sharp variations predicted in the model are not observed in the core and may be unlikely in a 3-dimensional environment with heterogeneous porosity and primary mineralogy.

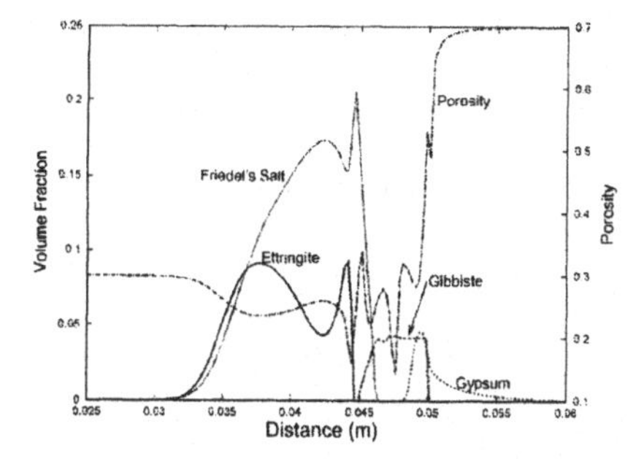

Figure 8: Calculated mineralogical and porosity profile across the carbonation zone for aluminous and sulfate-bearing minerals using case 32 at 30 years (Table 8). The initial cement-shale interface is at 0.05 m.

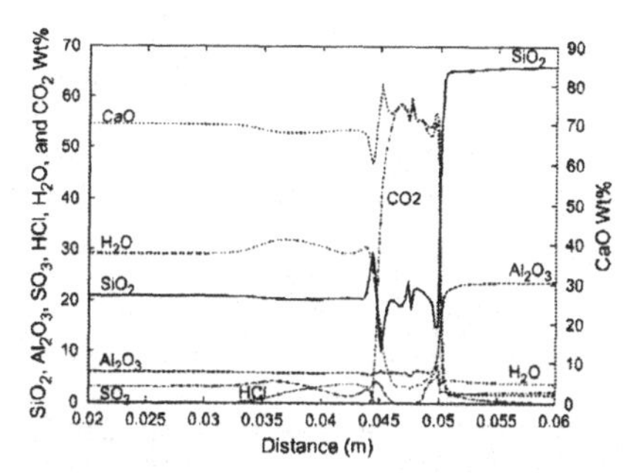

Figure 9: Calculated compositional profile across the carbonation zone for case 32 at 30 years (Table 8). The initial cement-shale interface is at 0.05 m. Compositional data for Na_2O, K_2O, and MgO were not modeled.

The numerical simulation provides predictions of detailed mineralogical features some of which are difficult to observe because of the amorphous character of the cement phases. At the reaction front (0.045 m), the simulation shows a rapid drop in pH that coincides with the precipitation of additional primary C-S-H ($X_{SiO_2} = 0.36$) as well as silica-enriched secondary C-S-H having a composition ranging from $X_{SiO_2} = 0.37$ to 0.44 (Figure 6 and 7). Interestingly, a dark-gray siliceous deposit separating the gray and orange carbonated zones occurs in the core and appears to be similar in character to the model prediction (Figure 5). There is then a very small spatial gap in the occurrence of secondary C-S-H followed by its reappearance at a much higher X_{SiO_2} ranging from 0.92 to 1.00 before it disappears just before the original cement-shale interface at 0.05 m. The occurrence of highly siliceous C-S-H coincides with the precipitation of amorphous silica and as the pH continues to drop, amorphous silica gradually replaces the secondary C-S-H.

The simulation suggests almost completely unaltered cement should exist > 2 cm from the original interface. We have not observed mineralogical changes within the lightly carbonated gray-colored cement in the core (e.g., occurrence of monosulfate or the occurrence of ettringite and Friedel's salt only adjacent to the carbonation zone). Some of this difference could be related to the difference between our model cement with monosulfate and the observation of very abundant hydrogarnet in the core.

In the model, infiltration of Cl^- from the brine attacks the monosulfate and forms Friedel's salt along with ettringite (Figure 8); both of these phases have been observed in the core samples. With further carbonation, the Friedel's salt and ettringite are predicted to break down to yield gypsum and gibbsite. The gibbsite in the model is an approximation for the amorphous Al-bearing material observed in the core. However, we have not observed gypsum in the carbonation zone; the predicted occurrence of gypsum is sensitive to species-dependent diffusion as discussed below.

The base case (Case 32) results were optimized to reproduce observations made on the core by selectively modifying key parameters. In the following section we consider the sensitivity of the carbonation model to the assumed porosity, solid solution model, tortuosity, and mineral reaction rates (Table 8). In the base case, we used a relatively high porosity for the shale (70%) compared to cement (30%) with the thought that the cement-shale interface was a porous region of CO_2 migration. This interpretation appears to be supported by numerical calculations that show using a porosity of the shale equal to (case 40) or less than the cement (case 42) results in parts of the reaction zone having little or no SiO_2 or Al_2O_3 (Figure 10) contrary to observations (Figure 5). The compositional gap in cases 40 and 42 is reflected in a mineralogical gap between the loss of the primary cement phases and the appearance of amorphous silica and gibbsite (Figure 11). However, the model parameters for case 42 do successfully reproduce the lack of a gypsum precipitate.

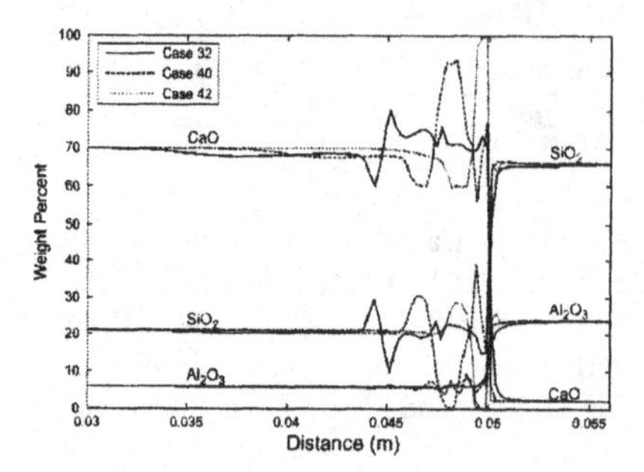

Figure 10: Calculated compositional profile across the carbonation zone for two different porosity distributions: Case 32: cement = 30% and shale = 70%; Case 40: cement = 30% and shale = 30%; and Case 42: cement = 30% and shale = 9%. The initial cement-shale interface is at 0.05 m.

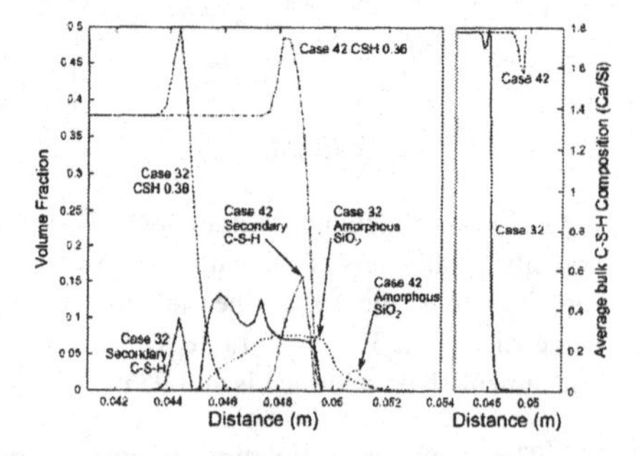

Figure 11: The distribution of the primary CSH phase ($X_{SiO_2} = 0.36$), secondary C-S-H, and amorphous silica for two different porosity distributions: Case 32: cement = 30% and shale = 70%; case 42: cement=30% and shale = 9%. The average composition of the secondary phase is shown in the right-hand panel. The initial cement-shale interface is at 0.05 m.

As illustrated in Figure 11, there are one or two regions of secondary C-S-H precipitation depending on the relative porosity of cement and shale. All of the model cases produce some secondary C-S-H at the carbonation front, within the envelope of primary C-S-H. This secondary C-S-H is moderately decalcified (to a minimum of Ca/Si $\approx$ 1.25; Figure 7). The amount of secondary C-S-H is strongly dependent on the porosity model such that the resulting average C-S-H composition is more silicic in the low shale porosity case (case 42; Figure 11). In many of the cases investigated, a second region of C-S-H precipitate developed within the carbonation zone (Figure 11). These compositions were always highly silicic (Ca/Si < 0.1). As these predicted compositions are not well constrained by experimental data (Figure 2), we believe there is little practical difference between highly silicic C-S-H and amorphous silica in the model.

Eliminating the C-S-H solid solution from the model does not significantly modify the width or geometry of the reaction zone (Figure 12). Instead, the amount of primary C-S-H ($X_{SiO_2} = 0.36$) at the reaction front increases to compensate for the absent secondary C-S-H; and the amount of amorphous silica within the carbonation zone increases to compensate for the absent highly silicic C-S-H. This difference in mineral distribution has little effect on the bulk composition (results not shown). Because we have not been able to determine the composition of the C-S-H in the recovered core, the determination of whether using the solid solution model more accurately captures the carbonation process is difficult. In part, this is a consequence of the nature of modeled environment: the CO_2-saturated brine that infiltrates the cement is out of equilibrium with any Ca-bearing C-S-H phase. As a consequence, the secondary Ca-rich C-S-H precipitates are destroyed along with the primary phase over a relatively narrow reaction front.

Figure 12 also illustrates the effect of suppressing both secondary C-S-H as well as amorphous silica precipitation (Case 37). This results in a very large increase in deposition of the primary C-S-H phase at the reaction front.

The results for differing porosity models shown in Figures 10 and 11 were initially surprising because cement grouts have porosity that are typically much greater than shales. The implications of the model results is that the interface between cement and country rock must have had a much higher effective porosity than the shale itself (i.e., the interface must have acted as a high porosity reservoir of CO_2-saturated brine). This is consistent with observational studies of the core[30]. A high porosity interface could have arisen from a relatively poor bond between the cement and shale or because of the presence of "wall cake" or ground material from the drilling process. The high porosity interface would provide a conduit for CO_2-saturated brine to move up from the CO_2 reservoir.

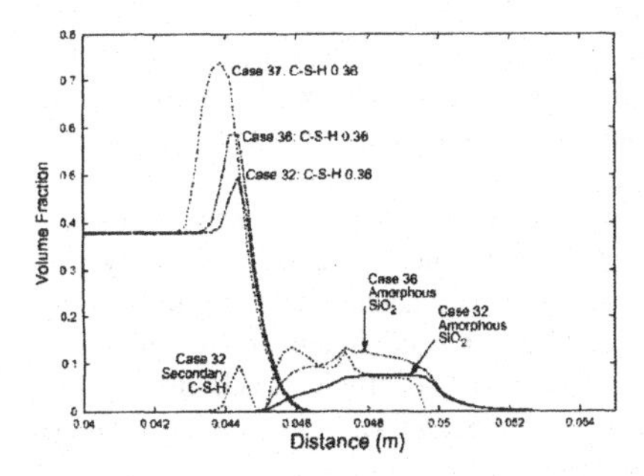

Figure 12: A comparison of model results using a solid solution model for C-S-H (Case 32), without a solid solution model (Case 36), and without a solid solution model or amorphous silica (Case 37). See Table 8.

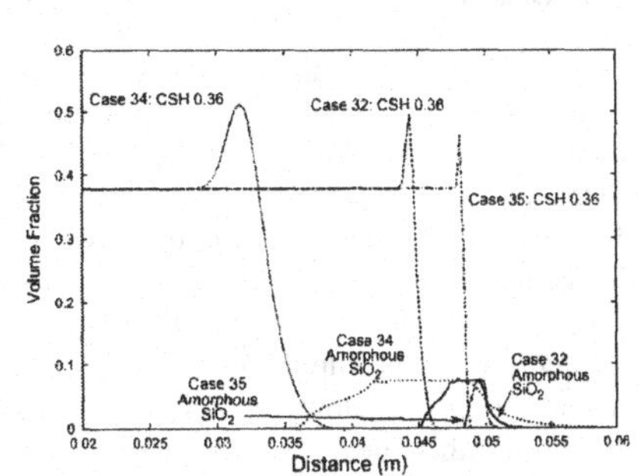

Figure 13: A comparison of model results using tortuosities differing by factors of 10 (see Table 8). The initial cement-shale interface is at 0.05 m.

Tortuosity limits the effective diffusivity of aqueous species and has a strong effect on the depth of penetration of the reaction (Figure 13). A $10\times$ larger tortuosity value (case 34; Table 8) moves the carbonation front $\approx$ 1 cm deeper into the cement and also widens the transition zone between the fully carbonated cement and the unaltered cement. Conversely, a $10\times$ smaller tortuosity value decreases and narrows the width of the carbonation zone. Differences in tortuosity values have little effect on the mineralogy of the reaction.

Decreasing the reaction rates of all of the phases by a factor of 100 eliminates almost all reaction (Case 38; Figure 14). Conversely, increasing the rates by a factor of 100 has little effect on the width of the reaction zone but allows the system to move closer to a local equilibrium approximation: amorphous silica replaces highly silicic secondary C-S-H and a distinct region of secondary C-S-H replaces the primary C-S-H at the reaction front.

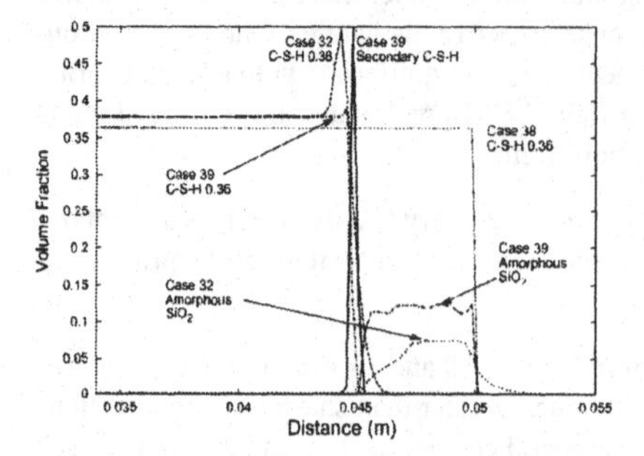

Figure 14: A comparison of model results with reaction rates scaled by factors of 100 (see Table 8). For Case 38, the amount of amorphous silica is < 0.01 and occurs between 0.025 and 0.05 m. Initial cement-shale interface is at 0.05 m.

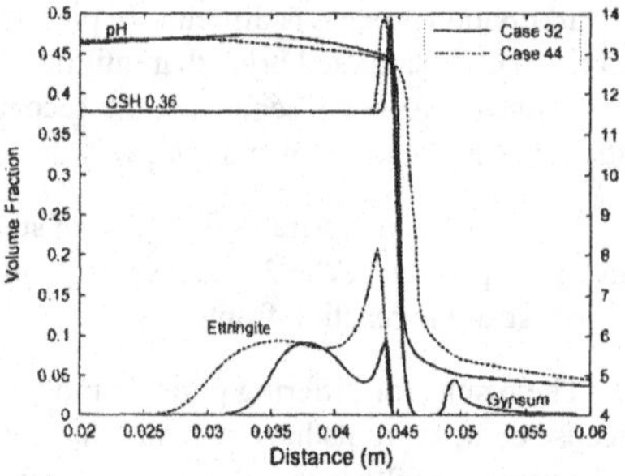

Figure 15: A comparison of a species-dependent aqueous diffusion coefficient calculation (Case 44) with a species-independent result (Case 32). Gypsum does not occur in Case 44.

In the previous examples, the diffusivities of all aqueous species were set equal. While this is computationally simpler, this simplification can hide subtle but important differences in the calculated mineral distribution. To explore this effect, we have used the species-dependent diffusion coefficients give in Table 4 with results shown in Figure 15. The coefficients range over a factor of 15. The results show shifts in the locations of phases along the 1-D column, partly driven by a change in the pH profile. Interestingly, the species-dependent calculation does not yield gypsum in agreement with observation. On the other hand, the species-dependent results suggest a much less uniform compositional profile across the reaction zone which is not in agreement with observation (not shown). However, note that the list of species-dependent coefficients is limited by lack of data–all other species are assumed to have the same coefficient value. As a consequence, the species-dependent calculations should be regarded as provisional.

Finally, we note that the use of activity coefficients formulated with the Pitzer model yielded only small changes in the position of reaction fronts and amount of reaction occurring in the cement (Figure 16). There were small changes in the composition of the secondary C-S-H where a greater range of compositions were precipitated (not shown). Use of the Pitzer equations results in an apparent decrease of the reaction zone width, which would require modification of the assumed tortuosity to match the observed 1-cm width of the carbonation zone.

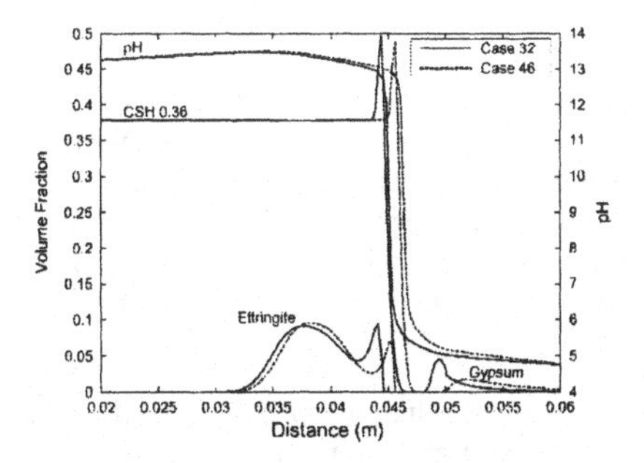

Figure 16: A comparison of a reactive transport results using extended Debye-Hückel activity coefficients (Case 32) and Pitzer activity coefficients (Case 46).

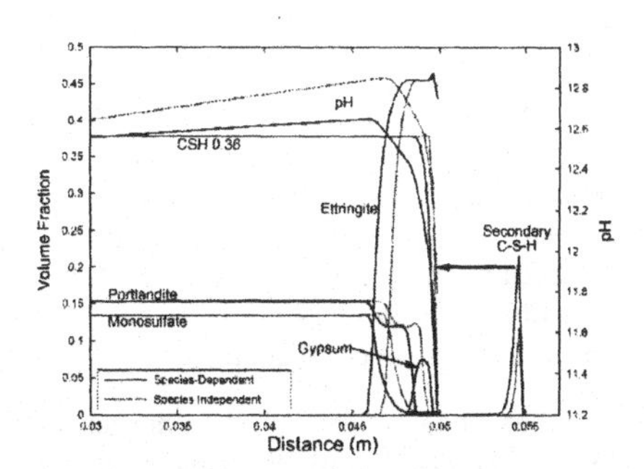

Figure 17: Calculated results of sulfate attack for both aqueous species-independent and aqueous species-dependent diffusion. The simulation domain was 0 - 0.05 m. The abundance of secondary C-S-H phases are shifted exactly +0.005 m for clarity. The average composition of the secondary phases ranges from X_{SiO_2} = 0.36 to 0.48. Gypsum does not occur in the species-independent case.

Our calculations show that the general features of carbonation reactions in oilwell cements can be reproduced with reactive transport calculations. In particular, the calculations show that decalcification reactions and subsequent precipitation of amorphous silica and alumina phases within a zone of intense carbonate precipitation are similar to field observations. The extent of reaction (depth of penetration) are critically dependent on the tortuosity and and to a lesser extent the mineral reaction rates. However, the simplified reaction rates used in this model are independent of the mineral surface area and more realistic models will require some form of pore-scale representation that captures both the available reactive surface area as well as the microstructural features that control tortuosity.

Sulfate Attack

Sodium sulfate attack is characterized by the conversion of monosulfate and other hydrated calcium aluminates to ettringite[32]. This process is associated with the loss of calcium hydroxide and decalcification of C-S-H. At more advanced stages in the process, gypsum also develops[33].

The sulfate attack problem was modeled with a different set of boundary conditions than the carbonation problem. In this 1-dimensional simulation, a 0.05 m section of cement represents a portion of a slab in contact with soil saturated with 0.05 molal Na_2SO_4 fluids. The soil contact is represented as a constant concentration condition, effectively simulating an unlimited reservoir of sulfate. The other boundary is a no-flow, constant gradient condition representing the unaffected interior of the slab. The phase abundance, phase properties, porosity, tortuosity are as in Case 32 (species-independent diffusion) or Case 44 (species-dependent diffusion) and Tables 6-8 as was the simulation time (30 years) and temperature (25 °C).

The depth of penetration of the sulfate reaction was significantly less than in the carbonation problem (Figure 17). The C-S-H phase is decalcified by sulfate ingress but is not lost from the reaction zone (amorphous silica does not occur). The most siliceous secondary C-S-H that precipitates is $X_{SiO_2} \approx 0.44$ (species-independent) and 0.49 (species-dependent). The monosulfate is replaced by a large amount of ettringite which results in a substantial reduction of the available porosity. Gypsum does not occur in the species-independent calculation but does occur in the species dependent calculation.

The model results reproduce the general features of sodium sulfate attack as described earlier. At present, we lack experimental or field results to perform a more complete evaluation along the lines of Maltais et al.[34]. However, similar to this work we also found that species-dependent diffusion coefficients were required to reproduce the precipitation of gypsum. Our model does provide a reasonable prediction of decalcification of C-S-H, the loss of calcium hydroxide, and the growth of ettringite.

CONCLUSIONS

The reactive transport code FLOTRAN was used to model carbonation and sulfate degradation processes in cement systems. The calculations were focused on exploring the role and significance of solid-solution in C-S-H and made use of the recent theoretical developments of Lichtner and Carey[13] that allow explicit treatment of compositional variation in solid solution systems within reactive transport calculations. In addition, the roles of porosity, tortuosity, species-dependent diffusion, and high ionic strength on predicted reaction pathways were examined.

A new thermodynamic model for solid solution in C-S-H was developed that approximated a full range of potential compositions between $Ca(OH)_2$ and SiO_2 endmembers. The model has a simple form and was based on the experimental data of Chen et al.[1] The model does not attempt to distinguish among the distinct compositional trends identified by Chen et al., but rather provides an estimate of an average composition. Although not explored here, the approach could be readily extended to model each of the distinct trends, all of which could be incorporated in the reactive transport calculations.

A brief overview of FLOTRAN provided key equations and approaches to incorporating solid solutions in reactive transport calculations. In addition, the methods used in FLOTRAN for handling species-dependent diffusion (which requires the introduction of electro-neutrality conditions and the Nernst-Planck equation) and high ionic strength fluids (based on aqueous species activity coefficients calculated with Pitzer equations) were introduced.

The cement degradation model was applied to carbonation of oilwell cement in a high-pressure (180 bar) environment and compared to observations of recovered cement core. The simulations were capable of reproducing the basic features of the observed carbonation reactions including the width of the alteration zone, phase distribution, and bulk composition. The model demonstrated that the porosity of the cement had to be less than the rock system supplying CO_2-saturated brine, consistent with observations of a disrupted interface between the oilwell cement and the shale country rock.

In the oilwell environment, CO_2-saturated brines are far from equilibrium with hydrated cement and their infiltration into cement leaves nothing but an amorphous residue and calcium carbonate. In this environment, the calculated solid solution behavior of C-S-H provides only a limited buffer against the infiltrating fluids. The solid solution model successfully predicts progressive decalcification of C-S-H and ultimate decomposition to calcium carbonate and either amorphous silica and/or a very Si-rich C-S-H. The amount of decalcification before decomposition was a function of the porosity ratio of cement and country rock as well as tortuosity with the amount of secondary C-S-H decreasing as the pH front encroached on the C-S-H front.

The model of sodium sulfate degradation showed monosulfate reacting to produce abundant ettringite and leading to a significant reduction in porosity. (Such a porosity reduction could lead to stress on the system through crystallization pressure.) Because of the less aggressive nature of the simulated sulfate attack, C-S-H persists at the interface with sulfate-bearing fluids and responds by decalcification.

The use of species-dependent diffusion coefficients had a particularly strong affect on the occurrence of gypsum. In the carbonation model, this eliminated gypsum from the alteration zone, which is consistent with observations of the core; and in the sulfate model, it produced gypsum, again consistent with observed sulfate attack processes. However, because the species-dependent diffusion coefficient database is rather incomplete, it is difficult to determine whether all features of cement degradation are more accurately characterized by these calculations. In fact, the bulk composition of the alteration zone resulting from the species-dependent calculation was less consistent with observations.

The use of Pitzer equations for aqueous activity coefficients produced no significant changes in the carbonation simulation. This occurred despite the use of an ≈ 1.6 molal brine.

The modeling results demonstrate that many features of cement degradation can be successfully reproduced. However, the more difficult problem of making predictions of the long-term behavior remains elusive. Principal uncertainties remain in the pore-scale geometries that govern the exposed surface area of reactive minerals (and thus the effective reaction rate of minerals) and the tortuosity. At present, there is no method of directly measuring the reactive surface area of individual phases within a cement sample and thus the value of the kinetic rate constants must be constrained by numerical modeling such as illustrated here. Although tortuosity can in principal be measured, methods to predict its changing value as mineral dissolution/precipitation occurs are still needed.

ACKNOWLEDGEMENTS

This work was supported by by the U.S. Department of Energy–NETL Contract #05FE01 and the Los Alamos Laboratory Directed Research and Development project DR 20030091DR. Kinder Morgan CO_2 Company is gratefully acknowledged for acquiring the core samples used in this study.

REFERENCES

[1] J. J. Chen, J. J. Thomas, H. F. W. Taylor, and H. M. Jennings. Solubility and structure of calcium silicate hydrate. *Cement and Concrete Research*, 34:1499–1519, 2004.

[2] E. J. Reardon. An ion interaction model for the determination of chemical equilibria in cement/water systems. *Cement and Concrete Research*, 20:175–192, 1990.

[3] E. J. Reardon. Problems and approaches to the prediction of the chemical composition in cement/water systems. *Waste Management*, 12:221–239, 1992.

[4] M. Kersten. Aqueous solubility diagrams for cementitious waste stabilization systems. 1. The C-S-H solid-solution system. *Environmental Science & Technology*, 30:2286–2293, 1996.

[5] J. J. Thomas and H. M. Jennings. Free-energy-based model of chemical equilibria in the CaO-SiO_2-H_2O system. *Journal of the American Ceramic Society*, 81:606–612, 1998.

[6] E. J. Reardon and P. Dewaele. Chemical model for the carbonation of a grout/water slurry. *Journal of the American Ceramic Society*, 73:1681–1690, 1990.

[7] D. A. Kulik and M. Kersten. Aqueous solubility diagrams for cementitious waste stabilization systems. 4. A carbonation model for Zn-doped calcium silicate hydrate by Gibbs energy minimization. *Environmental Science & Technology*, 36:2926–2931, 2002.

[8] E. M. Gartner and H. M. Jennings. Thermodynamics of calcium silicate hydrates and their solutions. *Journal of the American Ceramic Society*, 70:743–749, 1987.

[9] H. M. Jennings. Aqueous solubility relationships for two types of calcium silicate hydrate. *Journal of the American Ceramic Society*, 69:614–618, 1986.

[10] P. D. Glynn and E. J. Reardon. Solid-solution aqueous-solution equilibria: thermodynamic theory and representation. *American Journal of Science*, 290:164–201, 1990.

[11] P. D. Glynn, E. J. Reardon, L. N. Plummer, and E. Busenberg. Reaction paths and equilibrium end-points in solid-solution aqueous-solution systems. *Geochimica Cosmochimica Acta*, 54: 267–282, 1992.

[12] J. Doherty. PEST: Model-independent parameter estimation. Unpublished, issued by Watermark Numerical Computing, 2002.

[13] P. C. Lichtner and J. W. Carey. Incorporating solid solutions in geochemical reactive transport equations using a kinetic discrete-composition approach. *Geochimica Cosmochimica Acta*, 70:1356–1378, 2006.

[14] P. C. Lichtner, C. I. Steefel E. H., and Oelkers, editors. *Reactive Transport in Porous Media*, volume 34 of *Reviews in Mineralogy*. Mineralogical Society of America, 1996.

[15] P.C. Lichtner, S.B. Yabusaki, Pruess K., and C.I. Steefel. Role of competitive cation exchange on chromatographic displacement of cesium in the vadose zone beneath the Hanford S/SX tank farm. *Vadose Zone Journal*, 3:203–219, 2004.

[16] C.I. Steefel, D.J. DePaolo, and P.C. Lichtner. Reactive transport modeling: An essential tool and a new research approach for the earth sciences. *Earth and Planetary Science Letters*, 240:539–558, 2005.

[17] P. C. Lichtner. FLOTRAN user manual. Technical Report LA-UR-01-2349, Los Alamos National Laboratory, 2001.

[18] A.R. Felmy. GMIN, a computerized chemical equilibrium program using a constrained minimization of the Gibbs free energy: Summary report. *Soil Science Society of America, Special Publication*, 42:377–407, 1995.

[19] P.C. Lichtner and A.R. Felmy. Estimation of Hanford SX tank waste composition from historically derived inventories. *Computers in Geoscience*, 29:371–383, 2003.

[20] P.C. Lichtner. Critique of dual continuum formulations of multicomponent reactive transport in fractured porous media. In B. Faybishenko, editor, *Dynamics of Fluids in Fractured Rock*, volume 122, pages 281–298, 2000.

[21] A.R. Felmy and J.H. Weare. Calculation of multicomponent ionic diffusion from zero to high concentration: I. the system Na-K-Ca-Mg-Cl-SO_4-H_2O at 25°C. *Geochimica et Cosmochimica Acta*, 55:113–131, 1991.

[22] A.R. Felmy and J.H. Weare. Calculation of multicomponent ionic diffusion from zero to high concentration: II. inclusion of associated ion species. *Geochimica et Cosmochimica Acta*, 55: 133–144, 1991.

[23] J.S. Newman. *Electrochemical Systems*. Prentice Hall, 1973.

[24] R. Hasse. *Thermodynamics of Irreversible Processes*. Dover, 1969.

[25] E. Samson, J. Marchand, and Y. Maltais. Modeling ionic difffusion mechanisms in saturated cement-based materials—an overview. In R. Hooton, M. Thomas, J. Marchand, and J. Beaudoin, editors, *Materials Science of Concrete*, volume Ion and Mass Transport in Cement-Based Materials, pages 97–111, 2001.

[26] P. C. Lichtner. Continuum model for simultaneous chemical reactions and mass transport in hydrothermal systems. *Geochimica et Cosmochimica Acta*, 49:779–800, 1985.

[27] K.S. Pitzer. Characteristics of very concentrated aqueous solutions. In *Physics and Chemistry of the Earth*, volume 13 of *Physics and Chemistry of the Earth*, pages 249–272, 1981.

[28] P.C. Lichtner. The quasi-stationary state approximation to coupled mass transport and fluid-rock interaction in a porous media. *Geochimica et Cosmochimica Acta*, 52:143–165, 1988.

[29] B. Metz, O. Davidson, H. de Coninck, M. Loos, and L. Meyer. *IPCC Special Report on Carbon dioxide Capture and Storage*. Intergovernmental Panel on Climate Change, 2005. http://www.ipcc.ch/activity/srccs/index.htm.

[30] J. W. Carey, S. Wehner, M. Hirl, M. Raines, P. C. Lichtner, R. Pawar, and G. D. Guthrie Jr. Analysis of 30 year CO_2-brine interaction with a West Texas oil-well completion. In *Fourth Annual Conference on Carbon Capture & Sequestration, May, 2005, Alexandria, VA*, 2005.

[31] NETL. Comprehensive national brine database. Technical report, National Energy and Technology Laboratory, 2003.

[32] H. F. W. Taylor. *Cement Chemistry*. Academic Press, London, 1990.

[33] E. F. Irassar, V. L. Bonavetti, and M. González. Microstructural study of sulfate attack on ordinary and limestone Portland cements at ambient temperature. *Cement and Concrete Research*, 33:31–41, 2003.

[34] Y. Maltais, E. Samson, and J. Marchand. Predicting the durability of Portland cement systems in aggressive environments–laboratory validation. *Cement and Concrete Research*, 34:1579–1589, 2004.

CHEMO-PHYSICAL AND MECHANICAL APPROACH TO PERFORMANCE ASSESSMENT OF STRUCTURAL CONCRETE AND SOIL FOUNDATION

Koichi MAEKAWA
Department of Civil Engineering, University of Tokyo
7-3-1 Hongo,
Bunkyo-ku, Tokyo, 113-8656, Japan

Kenichiro NAKARAI
Department of Civil Engineering, University of Tokyo
7-3-1 Hongo,
Bunkyo-ku, Tokyo, 113-8656, Japan

Tetsuya ISHIDA
Department of Civil Engineering, University of Tokyo
7-3-1 Hongo,
Bunkyo-ku, Tokyo, 113-8656, Japan

ABSTRACT

Coupled analysis of mass transport and damage mechanics associated with steel corrosion and ASR is presented for structural performance assessment of reinforced concrete. Multi-scale modeling of micro-pore formation and transport phenomena of moisture and ions are mutually linked for predicting the corrosion of reinforcement and volumetric changes. The interaction of crack propagation with corroded gel migration is simulated. Two computer codes for multi-chemo physical simulation (*DuCOM*) and nonlinear dynamic mechanics of structural concrete (*COM3*) were combined. This computational system was verified by the laboratory scale experiments of damaged reinforced concrete members under static loads, and has been applied to safety and serviceability assessment of existing bridges. The coupled system is extended to soil foundation-underground water-reinforced concrete interaction of mechanics and mass transport.

INTRODUCTION – GENERAL SCHEME

In the scheme of performance-based design with more transparency to clients and taxpayers, performance assessment methods occupy a central position from a viewpoint of structural mechanics. This rational way of assuring the overall quality of infrastructures may create cost-beneficial design and construction that exactly satisfies several requirements assigned to engineers. Life-cycle performance of structures is being explicitly required and an appropriate design of materials and structures is sought. Furthermore, needs to verify remaining functionality of damaged existing facilities is rising for extending service life. To meet these challenges, keenly expected is an explicit prediction and simulation of structural life serviceability and safety under specified loads and ambient conditions. In this paper, the authors propose an integrated platform of solid mechanics and thermo-hydro dynamics of materials and structures with multi-scales of referential control volume on which each chemo-physics is applied. The constitutive model of cracked RC domain is discussed with regard to cracking, and overlay of thermo-hydro state variables is presented for multi-scale coupling with soil foundation.

MULTI-DIRECTIONAL CRACK MECHANICS

A scheme of RC modeling used for an integrated platform of both safety and life-cycle assessment is shown in Fig. 1. Multi-directional cracking and its interaction are taken into account by the active crack approach[1]. All microscopic physical states (cracking, yielding, shear slippage, remaining stiffness of fractured materials) are inherently included in the constitutive modeling. The stress carrying mechanisms are composed of compression/tension parallel and normal to cracking and shear transfer. By the active crack method[1], the primary cracking of governing no linearity of structural concrete is identified and path-dependent parameters are renewed in each load step of time. The plastic localization of reinforcement is of importance for simulating largely deformed elements. The spatial averaging of local stress and strain along reinforcement is applied for structural analysis with finite elements as shown in Fig. 2. Since the local yield occurs at the crack location and the rest of domain remains elastic, the averaged stress strain relation of reinforcement differs from that of a single bare bar. The following hardening of the element is much associated with extension of plastic zones and the averaged hardening stiffness is computed by considering the reinforcement ratio, tensile strength of concrete and properties of reinforcing bars[1].

When the load reversal is produced in a single direction, near orthogonal two ways cracking is experienced. Here, the crack-to-crack mutual interaction is not so great as to consider the shear transfer of each intersecting cracks. Then, the rotating crack methods and other models that assume coaxiality of stress and strain field function successfully for structural analysis, and the model of shear transfer does not play a central role of mechanics. However, the multi-directional and non-proportional loadings may create three and four directional cracking that intersects each other in finite element domain. When thermal and drying expansion and shrinkage would be coupled with seismic loads, principal stress directions drastically rotate. This general situation tends to create multi-directionally intersecting cracking with strong interaction. Figure 3 shows an example of experimental verification with three and four directional cracking in two-way reinforced RC panels under combined in-plane shear and normal stresses. The in-plane stresses were actively controlled by the internal hydraulic pressure, torsion moment produced by a couple of jacks and axial compression.

Constitutive models have to be verified on member/structural levels, because stress states and loading paths cannot be fully reproduced only by experiments at the element level. Shear wall experiments have been used for verification of in-plane RC modeling under monotonic as well as cyclic loads[1]. It is recognized that in-plane RC models are well applied under both static and dynamic excitation. Figure 4 also shows the experimental verification related to underground ducts and tunnels with and without soil foundation.

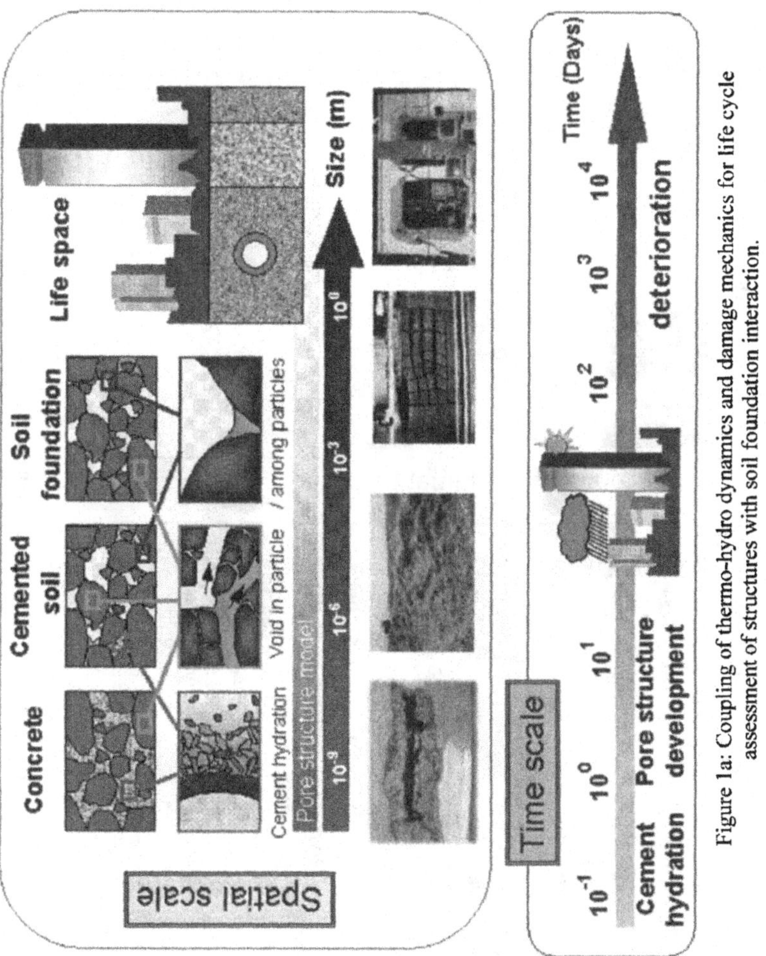

Figure 1a: Coupling of thermo-hydro dynamics and damage mechanics for life cycle assessment of structures with soil foundation interaction.

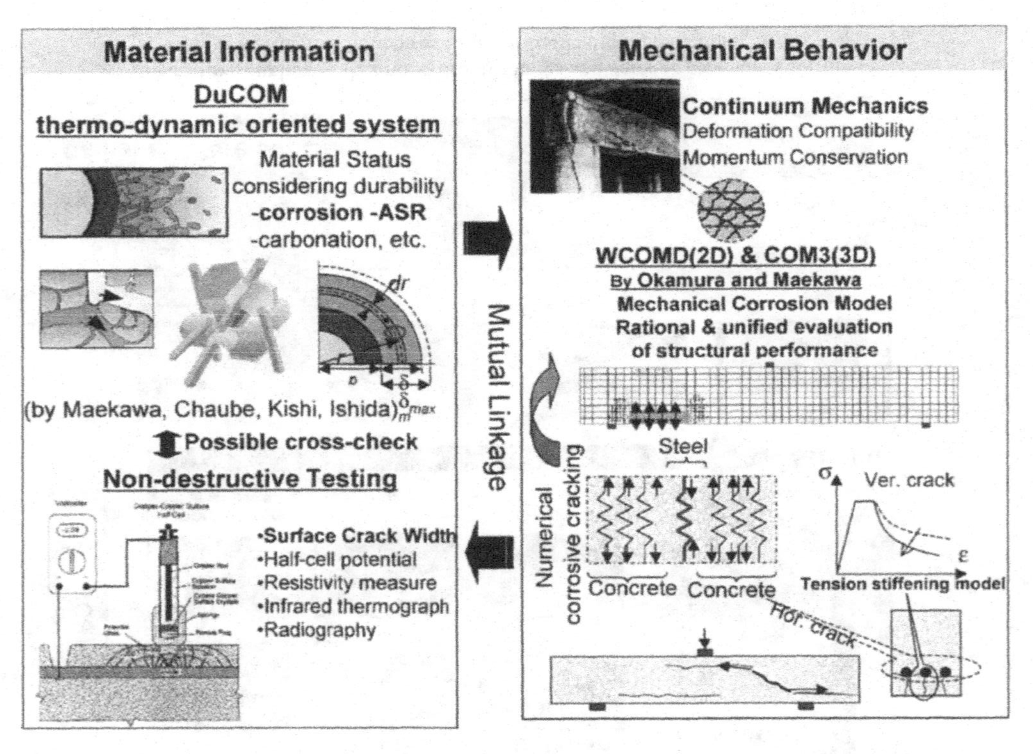

Figure 1b: Coupling of thermo-hydro dynamics and damage mechanics for life cycle assessment of structures with soil foundation interaction.

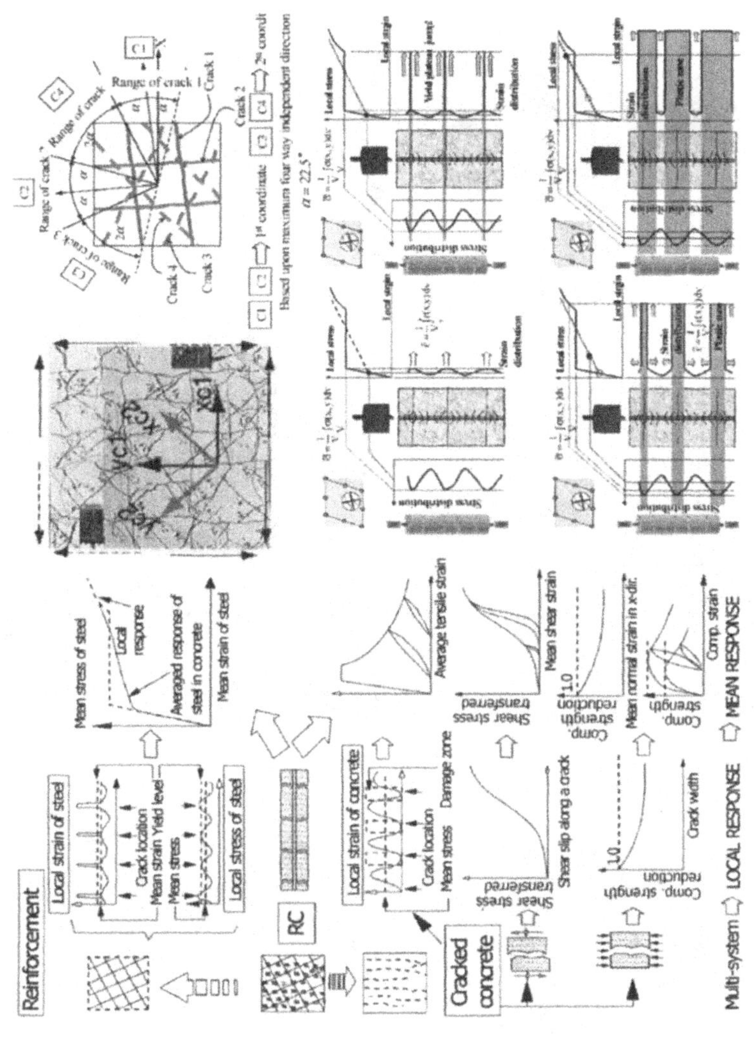

Figure 2: Formulation of in-plane constitutive model with multi-directional cracking[1]

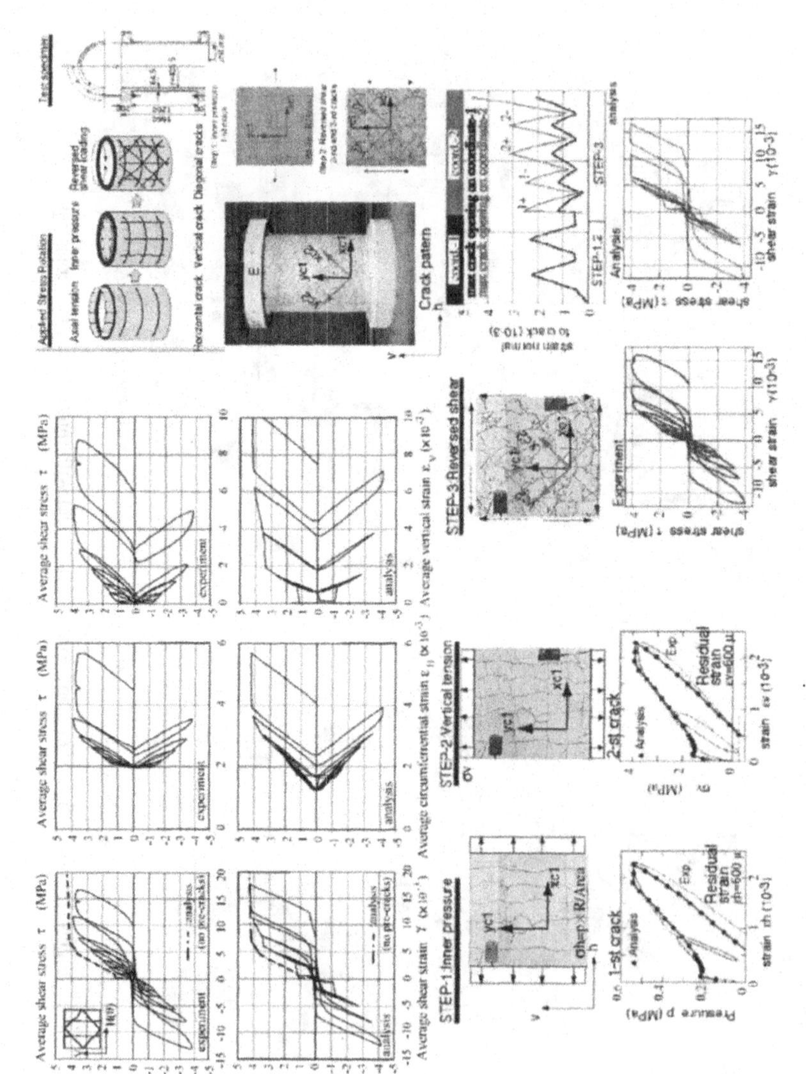

Figure 3: Experimental and analytical behavior of specimens A-2 (4-way cracks)[1]

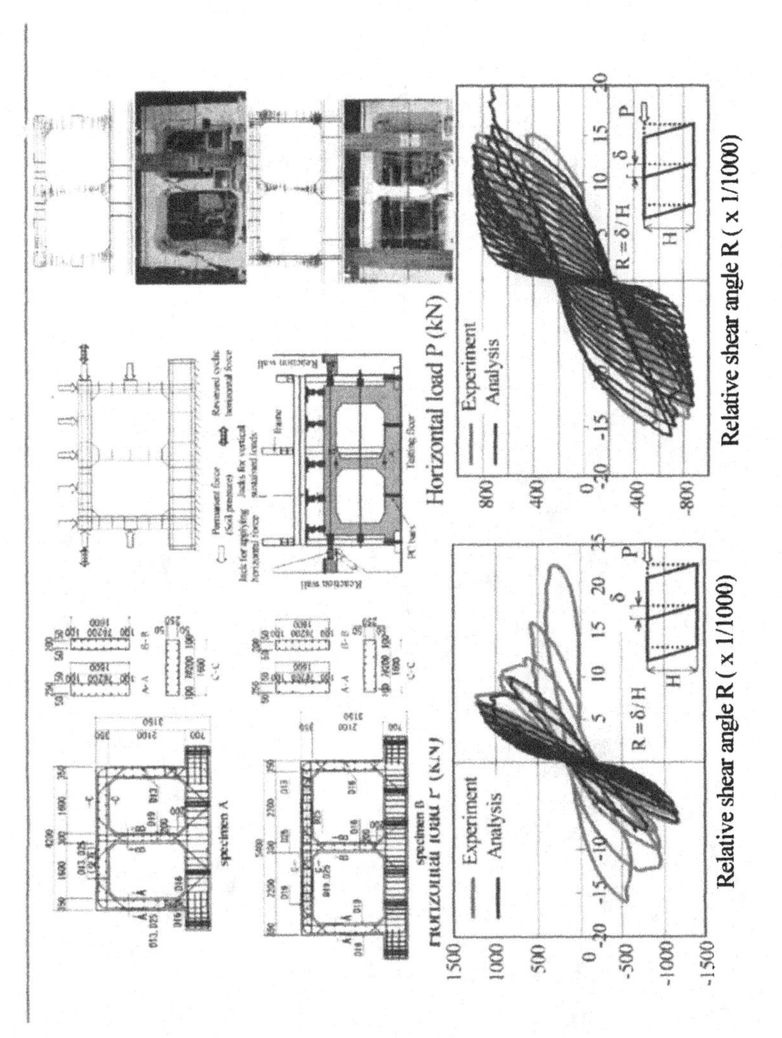

Figure 4a: RC underground box-vents subjected to static loads[1,4]

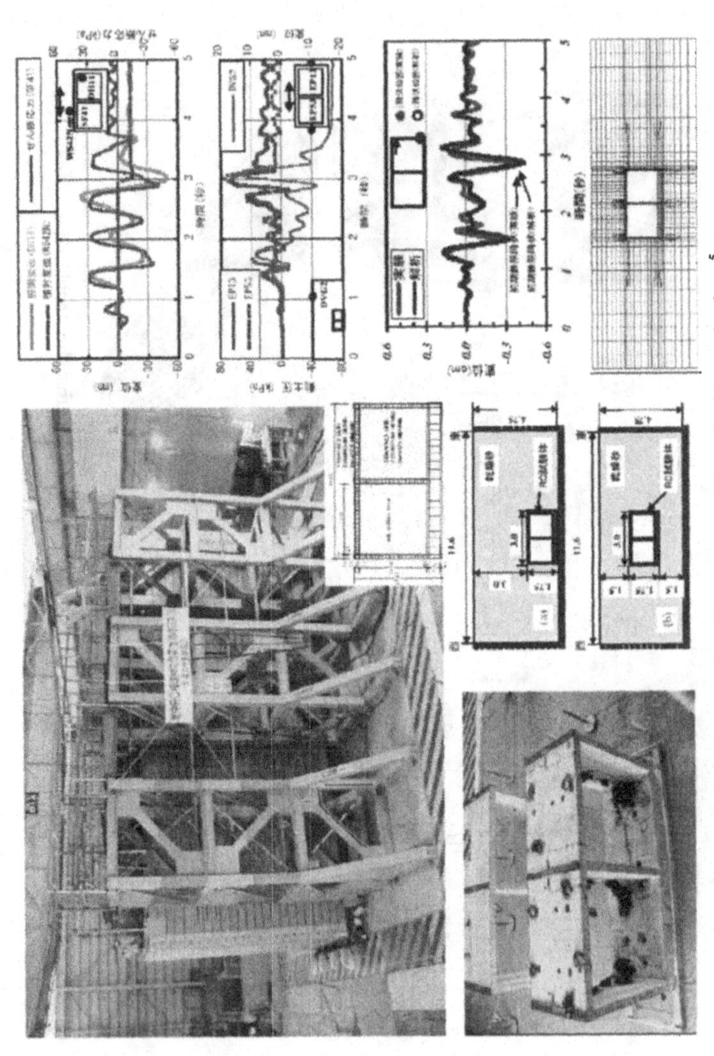

Figure 4b: RC underground box-vents subjected to dynamic shear[5]

THERMO-HYDRO CHEMO-PYSICAL MODELING

State variables of thermo-hydro dynamics are further required for life-cycle assessment. Volumetric change caused by temperature and long-term moisture equilibrium in micro-pores is associated with cracking and corresponding serviceability, and corrosion of reinforcement has

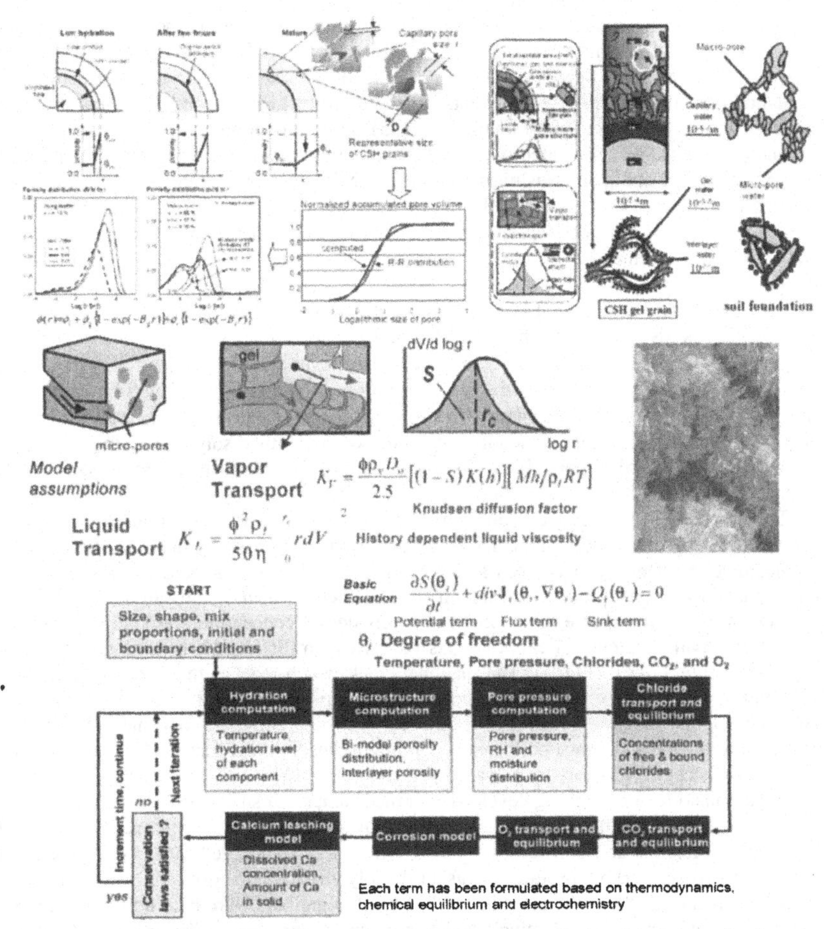

Figure 5: Micro-modeling of CSH gel and capillary pores and multi-chemo physics.

much to do with migration of chemicals through micro-pores. Thus, the coupled system as shown in Fig. 5 was proposed[2, 3] to simulate the entire thermo-mechanical states of constituent material and structures. For computing the thermo hydro equilibrium, multi-scale analysis platform *DuCOM*[2,3] was used. Micro-pore geometry and spaces are idealized by statically formulated pore distribution and internal moisture balance is simultaneously solved with mass conservation requirement. The moisture migration and diffusivity are computed based on the micro-pore size distribution and the space of condensed water channel.

Chloride ion migration and other chemical reactions such as carbonation and calcium leaching are overlaid on this system[3,6]. The conductivity and diffusion characteristics for mass transport are calculated based upon computationally formed micro-pore structure. The computation of multi-chemo-physical events is carried out by means of the sequential processing with closed-loop predictor-corrector method[3] (Fig. 5). The temperature dependent volume change and concrete shrinkage associated with microclimate in CSH gel and capillary pores are considered as offset strain in constitutive modeling. Micro-corrosion rate is also computed by simulating migration of O_2-CO_2 gas and chloride ion[2], and the effect of corrosion is integrated in the structural analysis[7]. These thermodynamic state variables are incorporated into the constitutive modeling before cracking.

COUPLING OF DAMAGE MECHANICS AND CHEMO-PYISCAL SCHEME

Cracking is also influential in mass transport of gases and dissolved ions. These cracks through which ion substances can easily migrate are mutually linked with thermo-hydro dynamic analysis by the hierarchy type of multi-scale modeling as shown in Fig. 5. This simulation can be mainly used for life-cycle assessment. Cracking of concrete causes accelerated diffusion of chloride. It may allow deeper penetration of chloride and other substances. In the analysis, diffusivity of substances is regarded as a variable in terms of computed averaged strain of concrete elements.

The corroded steel produces volumetric swelling and results in internal self-equilibrated stress, which may lead to additional cracking around reinforcing bars. Figure 6 illustrates the way to amalgamate the damage mechanics and volume expansion of generated corrosion gels. The effect of corrosion gel product formation is considered in the constitutive modeling of reinforcement in the transverse direction. If the corrosion is concentrated around the anchorage zone of main reinforcement, its capacity gets reduced with different crack propagation pattern from those of sound ones[7] (Fig. 7). The diagonal crack which reaches the bending compression zone is initiated by the corrosion crack tip created along the longitudinal main reinforcement. Finally, the diagonal crack is driven to the beam support. Apparently, the concentrated corrosion is seen to deteriorate the anchorage performance of longitudinal reinforcement. The acceleration test of corrosion of steel in RC beam by galvanostatic charge also substantiated this simulation result.

When the corrosion cracking develops over the beam, shear safety performance differs from the non-damaged reference case[10]. Figure 8 shows load-displacement relations for RC non-damaged reference and corroded specimen, which was submerged into a sodium pond for accelerated corrosion. Here produced was uniformly distributed corrosion along the whole longitudinal steel of 2.1% as the mass loss. Main reinforcing bars were bent up 90 degree inside the anchorage zone. Thus, comparatively satisfactory anchorage capacity is expected. In this case, the stiffness of the beam is much reduced but the capacity is a bit increased. The bond loss in the shear span leads to retarded propagation of diagonal shear cracking and may elevate the shear

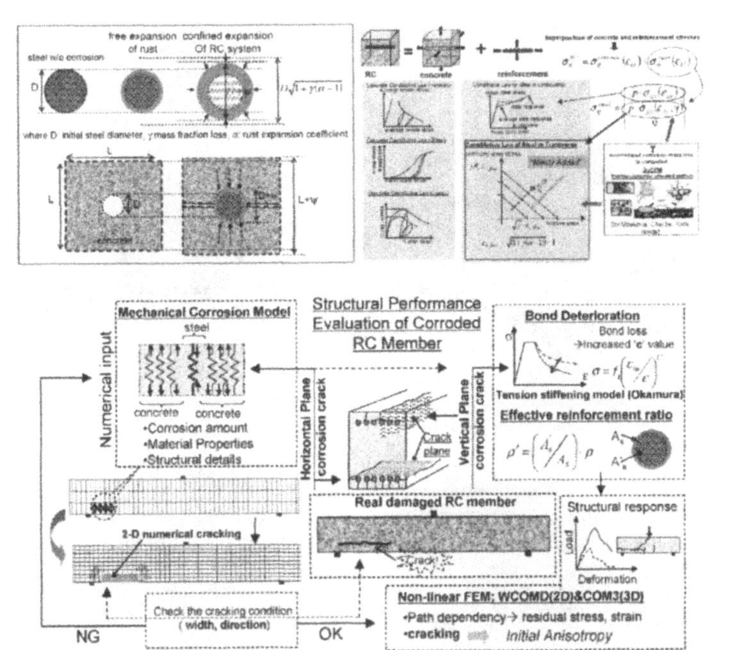

Figure 6: Simulation of corrosion of r/f bars and structural performance assessment[7]

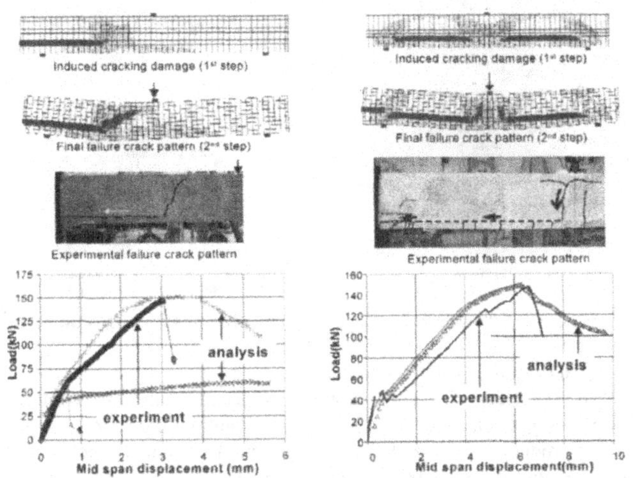

Figure 7: Analysis result of beam with inherent crack-like defect till the anchorage zone[7]

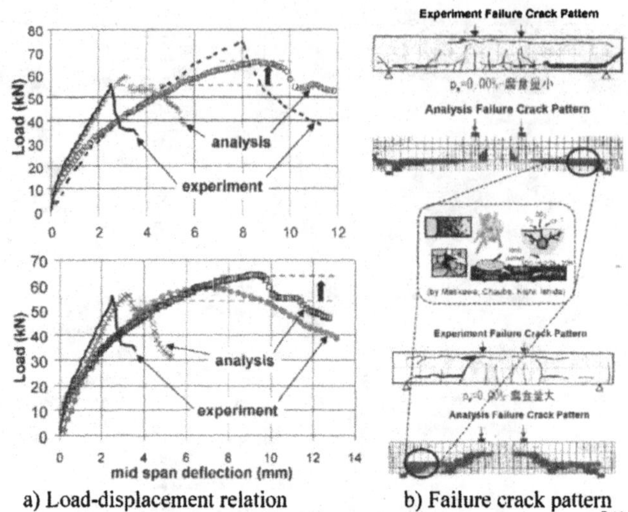

a) Load-displacement relation b) Failure crack pattern

Figure 8: Simulated shear capacity and cracking of corrosion beam[7,10]

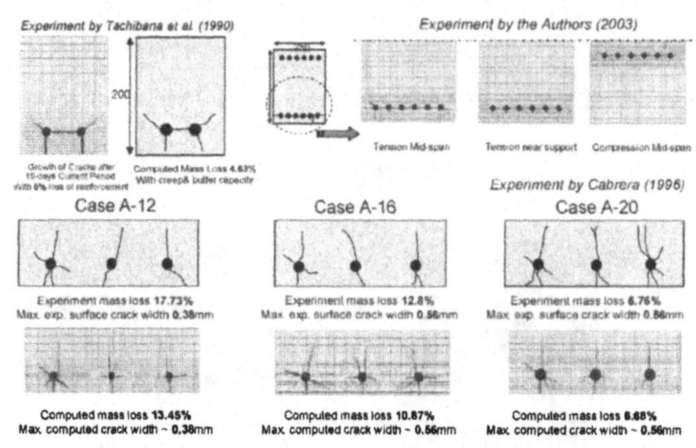

Figure 9: Simulated shear capacity and cracking of corrosion beam[8]

capacity. Computation can capture this property.

Figure 9 shows the corrosion crack propagation in experiment and simulation. The corrosive mass loss can be computed by *DuCOM* under the constant chloride concentration on the surface. The corrosion gel product is assumed to be created around the mother steel bars and the transverse stress normal to the reinforcement axis is computed as shown in Figure 6. The crack patterns of the same surface crack width and the corresponding corrosion mass loss are compared with each other.

The crack orientation and ligaments are fairly simulated. In these analyses, the injection of colloidal gels into crack gaps was taken into account in the analysis.

Figure 10 is the recent example of application to the engineering assessment for a 100-years old railway bridge in Tokyo[11]. Due to uneven settlement of the foundation, some initial damage remains in the form of cracking and arch ribs were strengthened by additional RC arch inside layer in the past. The seismic ground motion was applied to the numerically aged

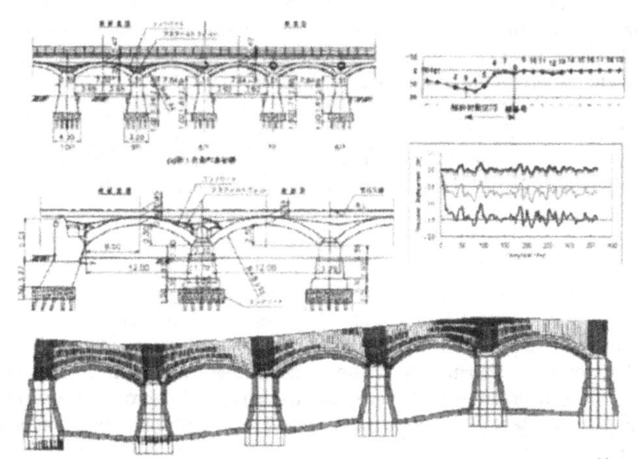

Figure 10: Safety assessment of 100 years-old railway bridges[11]

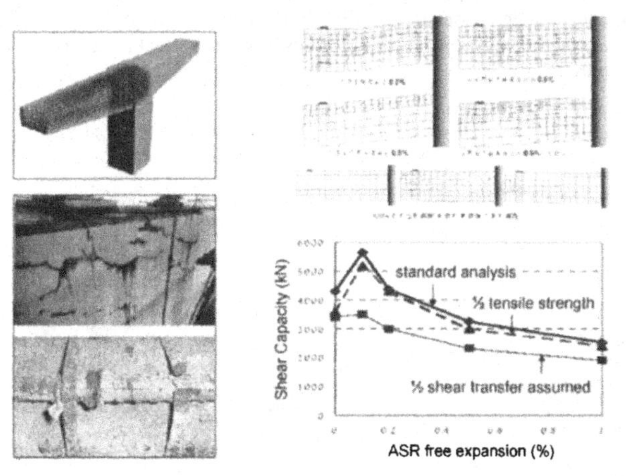

Figure 11: ASR damaged RC bridge pier and capacity simulation[12].

structural concrete and the computed response was used for safety and serviceability assessment in practice. The seismic remaining performance was numerically investigated and the sustainable life with light retrofit was judged.

Figure 11 shows the analysis for remaining structural safety of ASR damaged RC bridge piers. As lots of reinforcing bars are ruptured at the inside corners of bent portions, anchorage performance of web reinforcement is thought to be deteriorated. But, as a matter of fact, the capacity of ASR damaged members was predicted to increase in accordance withmagnitude of ASR expansion. This is attributed to the pre-stressing effect and self-equilibrated compressive axial force. After the peak of capacity, it starts to decline. Thus, the way of strengthening and/or repair must be different according to the induced expansion.

COUPLING WITH SOIL FOUNDATION SYSTEM

The multi-scale modeling of cementiitous composites can be extended to soil foundation with large-scale pores of strong connectivity. Figure 12 summarizes the micro-pore distribution modeling[6] and the mass transport modeling through CSH micro-pores was simply applied to the large-scale pores among soil grains. Verification was conducted in use of experimentally obtained permeability. Leaching and mass transport of calcium ion were mounted on this extended system for estimating extremely long-term performance of underground concrete structures and environmental issues of underground water. This extended simulation method can treat the life-cycle assessment of cementitious soil. Figure 13 shows analytical and experimental results of calcium leaching from the cemented sand. Leaching is associated with permeability and bulk motion of water, whose characteristic is greatly influenced by the micro-pore structure. This is not a given material constant but computed value of cementitious composite in the multi-physicochemical scheme.

Figure 14 shows an example analysis and verification of calcium leaching from underground concrete structures into the soil environment. Different boundary conditions are assumed; exposed to water with constant concentration and soil foundation with no bulk motion of underground water. The results indicate that the interaction with soil foundation is critical for assessment of leaching of underground structure (Fig. 14).

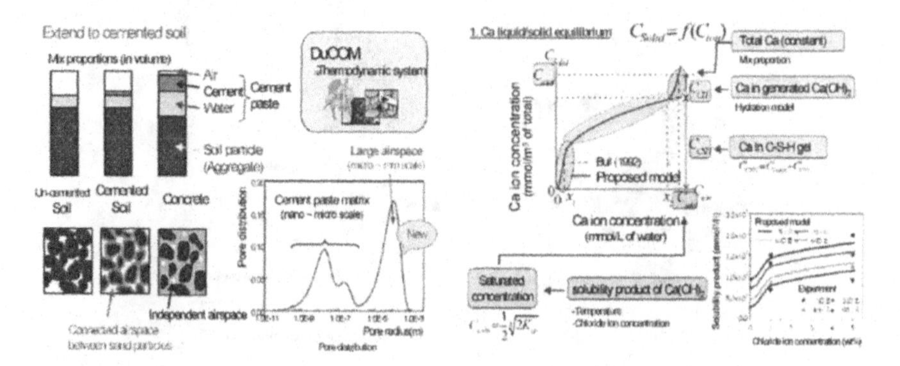

Figure 12: Extension of micro-pore model and calcium leaching from CSH to soils[6,13].

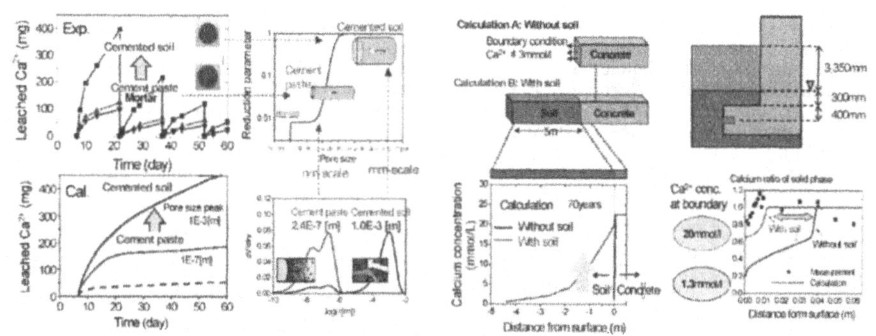

Figure 13: Calcium leaching of cemented soil[6] Figure 14: Verification of calcium leaching from underground concrete [6]

CONCLUSIONS

Chemo-physical and mechanical modeling of concrete with greatly different scales of geometry was presented, and synthesized on a unified computational platform, which may bring about quantitative assessment of structural concrete performances of interaction with soil foundation. The safety assessment method was extended to the life-cycle issue with multi-scaled information on microclimate states of cementitious composites. Currently granted is a great deal of knowledge earned by the past development. At the same time, we face a difficulty to quantitatively extract consequential figures from them. The authors expect that the systematic framework on the knowledge-based technology will be extended efficiently and can be steadily taken over by engineers in charge. This study was financially supported by Grant-in-Aid for Scientific Research (S) No.15106008.

REFERENCES

1. Maekawa, K., Pimanmas, A. and Okamura, H., "Nonlinear Mechanics of Reinforced Concrete", *Spon Press*, London (2001).
2. Maekawa, K., Chaube, R. P. and Kishi, T., "Modeling of Concrete Performance", *Spon Press*, London (1999).
3. Maekawa, K., Ishida, K. and Kishi, T., "Multi-scale modeling of concrete performance – integrated material and structural mechanics –", *Journal of Advanced Concrete Technology*, 1:2, 91-126 (2003).
4. Soraoka, H., Adachi, M., Honda, K. and Tanaka, K., "Experimental study on deformation performance of underground box culvert", *Proceedings of the JCI*, 23:3, 1123-1128 (2001).
5. Japan Society of Civil Engineers, "Recommendation for structural performance verification of LNG underground storage tanks", *Concrete Library* 98, JSCE, Japan (1999).
6. Nakarai, K., Ishida, T. and Maekawa, K., "Multi-phase physicochemical modeling of soil-cementitious material interaction", *Proceedings of JSCE*. No.802/V69. (2005).
7. Toongoenthong, K. and Maekawa, K., "Multi-Mechanical Approach to Structural Performance Assessment of Corroded RC Members in Shear", *Journal of Advanced Concrete Technology*, 3:1, 107-122 (2004).
8. Toongoenthong, K. and Maekawa, K., "Simulation of Coupled Corrosive Product Formation, Migration into Crack and its Propagation in Reinforced Concrete Sections", *Journal of Advanced Concrete Technology*, 3:2, 253-265, (2005).
9. Toongoenthong, K. and Maekawa, K., "Interaction of pre-induced damages along main reinforcement and diagonal shear in RC members", *Journal of Advanced Concrete Technology*, 2:3, 431-443 (2004).

[10] Satoh, Y. et al., "Shear behavior of RC member with corroded shear and longitudinal reinforcing steels", *Proceedings of the JCI*, 25:1, 821-826. (2003).

[11] Sogano, T. et al., "Numerical analysis on uneven settlement and seismic performance of Tokyo brick bridges", *Structural Engineering Design*, No.17, JR-East, 96-109 (2001).

[12] Jaoan Society of Civil Engineers, "Report on safety and serviceability of ASR damaged RC structures", *Concrete library of JSCE*, No.XX (2005).

[13] Nakarai, K., Ishida, T., Maekawa, K. and Nakane, S., "Calcium leaching modeling of strong coherency with micropore formation of porous media and ion phase equilibrium", *Proceedings of JSCE*. No.802/V69 (2005)

RESTRAINT AND CRACKING DURING NON-UNIFORM DRYING OF CEMENT COMPOSITES

John Bolander, Zhen Li and Mien Yip
Department of Civil and Environmental Engineering
University of California, Davis, USA

ABSTRACT

The drying of cement composites causes shrinkage that, when coupled with restraint, leads to stress production and possible cracking of the material. Shrinkage induced cracking is a primary concern, since most cracking fosters the ingress of harmful substances that can compromise the long-term durability of structural components. To reduce shrinkage cracking as a design objective, the various factors affecting the process should be understood within the context of the three-dimensional structural component and its boundary conditions. These factors include the non-uniform movement of moisture due to drying, the aging of the cement matrix, creep mechanisms, and several forms of restraint. This paper reports on results obtained using three-dimensional simulations of cement composites subjected to drying environments. Attention is given to the development of shrinkage cracking in statistically homogeneous models of concrete, as affected by the modeling of moisture diffusivity. Fundamental work on the role of material features on moisture transport is also reported. The numerical framework explicitly models inclusions and their interface with the surrounding matrix, offering the potential for more detailed simulations of moisture transport in concrete materials. Aspects of the simulation results are verified using theoretical models and through qualitative comparisons with experimental data.

INTRODUCTION

The cement paste component of concrete materials shrinks due to hydration and the loss of pore humidity to drying environments. In the presence of restraint, this shrinkage produces stress that can be large enough to cause cracking. There are various forms of restraint against free shrinkage, including: structural boundary conditions; composite material effects (e.g. due to mismatches of the mechanical and hygral properties of the material constituents); composite structural effects; and self-restraint activated by humidity or temperature gradients within a structural component. Humidity gradients can develop from non-uniform drying of a structural component. Exposed surfaces, from which moisture loss occurs, typically experience tension whereas the other regions react to preserve equilibrium. The mechanisms of drying shrinkage are commonly viewed from a materials science perspective, in which (at most) basic restraint conditions are assumed. In most practical situations, however, the various forms of restraint are present in combination and in varying degrees, depending on the interplay between material design, environmental exposure, curing conditions, structural dimensions, and structural boundary conditions. For completeness, it should be mentioned that concrete creeps and its properties also change with degree of maturity and ageing.

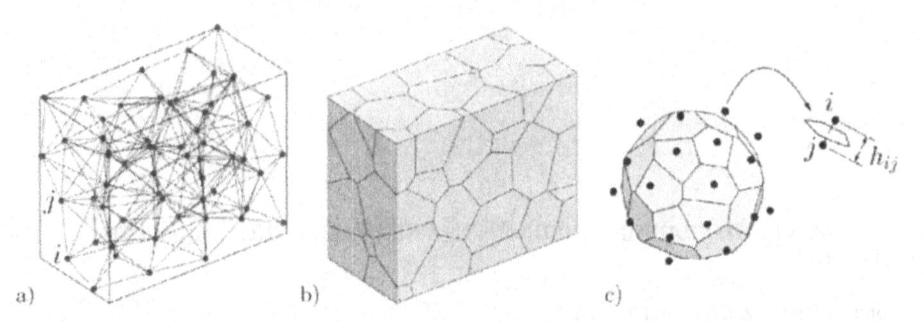

Figure 1 a) Lattice model topology (Delaunay tessellation); b) volume rendering (Voronoi tessellation); and c) inclusion discretization based on an irregular set of points

The life-cycle design of concrete structures should be supported by models that account for restrained shrinkage and other factors that affect cracking[1], both prior to and during service conditions. As crack opening and crack spacing is essential information for durability design, the models should account for the fracture mechanics of concrete materials[2]. Due to the range of length and time scales over which these problems exist, researchers have developed both: 1) macroscopic models, which treat concrete as a homogenous material (and represent fracture and drying through macroscopic notions of fracture energy consumption and moisture diffusivity, respectively); and 2) micro/mesoscale models that link the properties, distribution, and processing of material constituents to composite performance. Durability mechanics problems are almost exclusively three-dimensional in nature, especially when considering the above forms of restraint and the explicit modeling material features at the micro/mesoscales.

This paper reviews a lattice-type approach to modeling concrete materials, with emphasis on the macroscopic modeling of drying shrinkage cracking within elementary structural components. For a given set of material parameters, structural boundary conditions, and environmental exposure, the modeling of moisture diffusivity is shown to strongly influence the developing crack patterns. Attention is then directed toward the explicit modeling of aggregate inclusions and their local effects on moisture transport. The model is unique in that the interfacial transition zone (ITZ) between the inclusions and the matrix is represented by an irregular triangular lattice that runs parallel to the inclusion surface. A method is proposed for calculating nodal fluxes, which is then used to portray streamlines of moisture flow about various forms of a spherical inclusion.

MATERIAL MODELING

Research on coupled moisture transport/stress analyses using lattice models has been forwarded by Sadouki and van Mier[3]. In this work, moisture transport and elasticity of the material are modeled using an irregular lattice, such as that shown in Fig. 1. Both the scalar field of relative humidity and the vector field of generalized displacements are determined at the same set of lattice nodes. The following sections present basic aspects of the model. Additional details and verifications of model accuracy are given elsewhere[4,5,6].

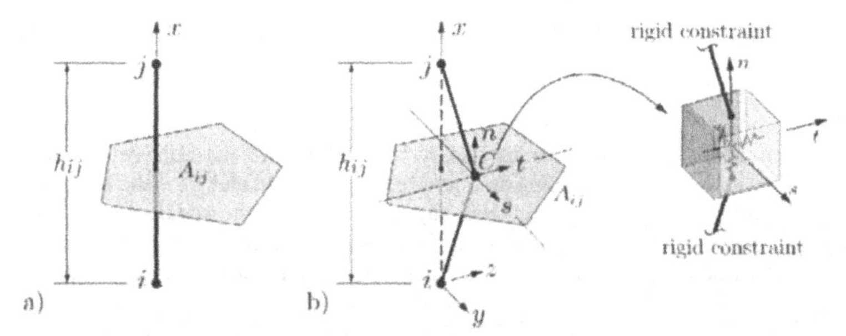

Figure 2 Lattice elements: a) conduit element; and b) rigid-body-spring element

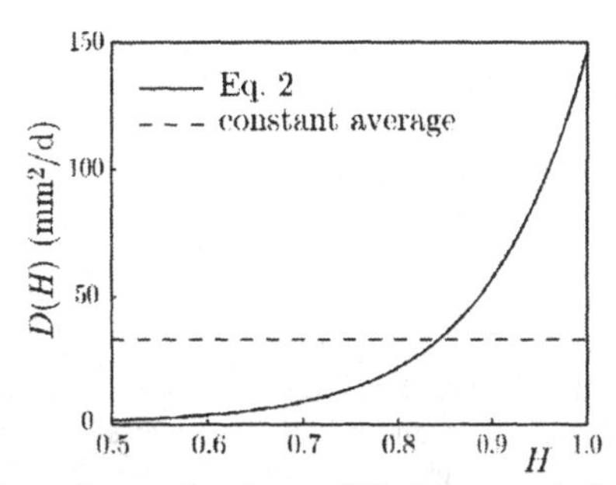

Figure 3 Dependency of moisture diffusivity on relative humidity

Lattice Model of Moisture Diffusion

The works of Bažant & Najjar[7] and Martinola & Wittmann[8] are used as a general framework for modeling moisture diffusion in cement composites, the convective boundary conditions at exposed surfaces, and the coupling of the diffusion and strain analyses. Assuming isothermal conditions, with no sinks or sources present, the governing equation for transient nonlinear diffusion is

$$\frac{\partial H}{\partial t} = \mathrm{div}[D(H)\mathrm{grad}H] \tag{1}$$

where H is the pore relative humidity and D is the diffusion coefficient, which generally depends on H. Equation 1 models moisture diffusion in a macroscopic sense and D should be regarded as an apparent moisture diffusion coefficient, representing complex processes and dependencies on the material constituents[9]. In this work, the diffusion coefficient is assumed to vary according to:

$$D(H) = \beta\exp(\gamma H) \tag{2}$$

where β and γ are parameters that can be determined by calibration with experimental results, as has been done for the exponential curve shown in Fig. 3[10]. Moisture exchange between the material boundary and the atmosphere is modeled using convective boundary conditions of the form

$$q_s = C_F(H_s - H_a) \tag{3}$$

where q_s is moisture flux across the boundary, H_s and H_a are the relative humidities at the material surface and in the surrounding atmosphere, respectively, and C_F is the film coefficient. The lattice topology is defined by the Delaunay tessellation of the nodal points[11], as shown in Fig. 1a. Each edge ij of the tessellation can be viewed as a conduit element[4] that transports moisture between nodes i and j (Fig. 1a). The elemental capacity and diffusivity matrices

$$\mathbf{M_e} = \frac{h_{ij}A_{ij}}{6d}\begin{bmatrix} 2 & 1 \\ 1 & 2 \end{bmatrix} \tag{4}$$

$$\mathbf{K_e} = \frac{DA_{ij}}{h_{ij}}\begin{bmatrix} 1 & -1 \\ -1 & 1 \end{bmatrix} \tag{5}$$

are assembled to form the corresponding system matrices, **M** and **K**. The semi-discrete form of Eq. 1 is then

$$\mathbf{M}\dot{\mathbf{H}} + \mathbf{K}\mathbf{H} = \mathbf{f} \tag{6}$$

where **H** is the vector of nodal relative humidities and the dot over **H** indicates time derivative. Element length h_{ij} is the distance between nodes i and j; cross-section area A_{ij} is set equal to the area of the Voronoi facet corresponding to the same two nodes (Fig. 2a). Apart from the Voronoi determination of A_{ij} and h_{ij}, this is essentially the same procedure as would be done for the finite element analysis of potential flow problems using two-node elements[12]. In the elemental capacity matrix, d = 1.0, 2.0, and 3.0 for 1-d, 2-d, and 3-d networks, respectively. A Crank-Nicolson algorithm is used to solve the diffusion equations, along with iterations that account for the dependence of the diffusion coefficient on H.

Lattice Model of Material Elasticity and Fracture

Frame-type lattice elements are used to model the elasticity of three-dimensional media. The basic element is composed of a zero-size spring set, located at the area centroid (point C) of the Voronoi facet common to nodes i and j, and rigid arm constraints that link the spring set with the nodal degrees of freedom (Fig. 2b). Each lattice node has six degrees of freedom. This approach is based on the rigid-body-spring concept of Kawai[13]. The spring set consists of three unidirectional springs, oriented normal and tangential to the facet, and three rotational springs (not shown in Fig. 2b) about the same local axes. The uniaxial springs are assigned equal stiffness values:

$$k_n = k_s = k_t = E\frac{A_{ij}}{h_{ij}} \tag{7}$$

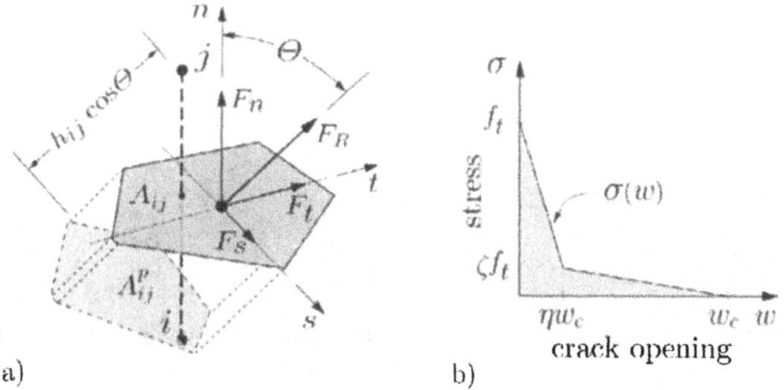

Figure 4 Crack band model of fracture: a) crack band constrained to Voronoi geometry; and b) crack softening relation

whereas the rotational springs are assigned stiffnesses:

$$k_{\phi n} = E\frac{J_p}{h_{ij}}; \qquad k_{\phi s} = E\frac{I_{ss}}{h_{ij}}; \qquad k_{\phi t} = E\frac{I_{tt}}{h_{ij}} \tag{8}$$

where E is the elastic modulus; J_p is the polar moment of inertia of the facet area; and I_{ss} and I_{tt} are the two principal moments of inertia of the facet area. The principal axes of the polygonal facet and its normal define the local n-s-t coordinate system shown in Fig. 2b. These basic components are used to construct the element stiffness matrix, as described by Berton and Bolander[6]. The use of Eqs. 7 and 8 (where h_{ij} and A_{ij} are defined by the Voronoi diagram) to assign the spring stiffness coefficients provides an elastically uniform lattice, i.e., a lattice that can reproduce uniform strain fields. This condition of elastic uniformity is prerequisite to introducing heterogeneity, such as aggregate inclusions.

In this paper, shrinkage strains in the elastic material arise from changes in relative humidity at the lattice nodes, as determined from the diffusion analyses described above. A constant hygral shrinkage coefficient, α_{sh}, is used in this coupling of the elasticity and diffusion analyses[8]

$$\Delta\epsilon_S = \alpha_{sh}\Delta H \tag{9}$$

where ΔH is the difference in nodal relative humidity over the computational time step. Equation 9 uses the average of ΔH calculated at the nodes of a given element and $\Delta\varepsilon_S$ is then introduced in the n (facet normal) direction of the corresponding element. This approach provides uniform volumetric straining of the lattice if ΔH is constant over the domain.

A crack band model[14] is used to simulate fracture within the lattice elements. This involves degrading the strength and stiffness of the element spring sets (Fig. 2b), in accordance with a prescribed softening relation (such as that shown in Fig. 4b, where σ is the traction acting across the crack and w is the crack opening displacement.) The softening parameters can be determined through inverse analyses of fracture test data[6]. Key features of this fracture model are that: 1) the crack band forms perpendicular to the principal tension direction, as determined by the resultant of the spring set forces, F_R (Fig. 4a). This direction generally differs from that of the element axis by angle θ; and 2) the dimensions of the crack band conform to the geometry of the

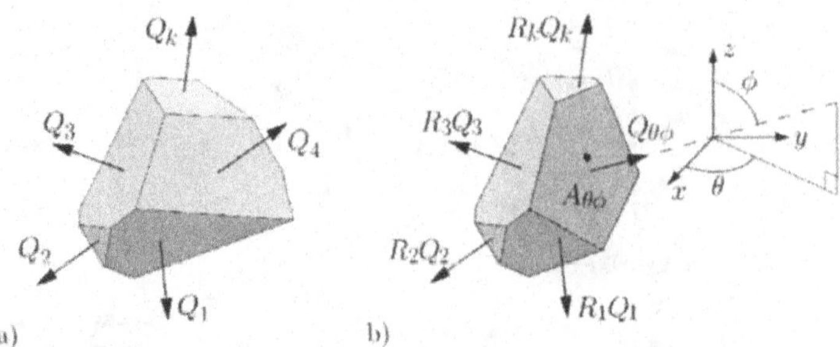

Figure 5 a) Elemental flow through a Voronoi cell; and b) sectioning of Voronoi cell for nodal flux calculation

Voronoi diagram (i.e. the crack band volume is the product of the crack band width, h_{ij} cosΘ, and the projected area of the Voronoi facet in the direction of tension, A_{ij}^P.) This approach provides energy conserving, grid insensitive representations of fracture within two- and three-dimensional lattices[5,6].

VISUALIZATION OF RESPONSE QUANTITIES

One primary application of lattice models is the discrete representation of material features at very fine scales, where the assumption of material continuity breaks down. When modeling at coarser length scales, the reduced dimensional (1-D) representation of material can still be advantageous, although there are difficulties in interpreting the lattice as a continuum. The effectiveness of the lattice as a modeling tool depends on the ability to interpret the analysis results, particularly for 3D problems. As described hereafter, the Voronoi discretization of the domain facilitates visualization of the model results in several ways, including the accurate calculation of nodal gradients of the field quantities and the depiction of cracking in three dimensions.

Nodal Flux Calculation

The lattice modeling moisture diffusion is analogous to other network models of potential flow, where potential is represented at the network nodes and flow is conveyed via edge variables[15]. In contrast to flow in discrete networks (e.g. electrical circuits), the lattice model represents flow within a continuum. However, the orientation of the individual lattice elements is not related to any material or structural feature and, therefore, the flow through the lattice elements does not have practical meaning. To determine the direction and magnitude of flow within the continuum, a procedure for calculating nodal flux has been developed. This procedure, presented hereafter, is an extension of previous work for 2D lattices[4].

After solving the governing field equations to determine nodal potentials, the element equations are used to determine the flow within each element. The elements, and therefore the flows, framing into a given node (or cell) are known (Fig. 5a). The net flow into the node is zero based on conservation of mass of the transport substance, which is analogous to Kirchhoff's current law for electrical circuits. To determine nodal flux in a given direction, defined in spherical coordinates by angles θ and ϕ, the Voronoi cell associated with node i is sectioned through the node and perpendicular to the direction of interest, as shown in Fig. 5b. The flow $Q_{\theta\phi}$

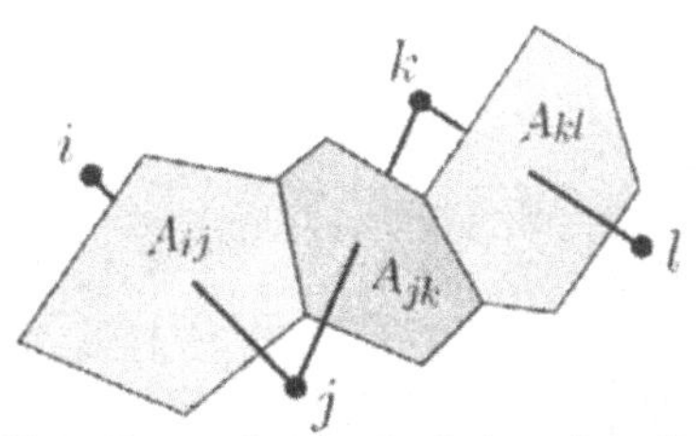

Figure 6 Piecewise continuous depiction of crack trajectory

through the newly established cut face is determined by summing the weighted flow contributions of all n elements framing into the node:

$$Q_{\theta\phi} = \sum_{k=1}^{n} R_k Q_k \tag{10}$$

The weighting factor $R_k = A_k'/A_k$, where A_k is the area of facet k and A_k' is the area of facet k on the negative side of the cut plane, in terms of the normal direction (θ, ϕ). For facets intersected by the cut plane, $0 < R_k < 1$; otherwise, R_k equal zero or unity depending on whether all vertices of the facet are situated on the positive or negative side of the cut plane, respectively. The flux is then $q_{\theta\phi} = Q_{\theta\phi} / A_{\theta\phi}$ where $A_{\theta\phi}$ is the area of the cut face (Fig. 5b). To determine flow vectors, $q_{\theta\phi}$ is calculated for a discrete, regular sampling of all orientations of the cut face. The orientation with maximum $q_{\theta\phi}$ defines the flux direction and magnitude. This procedure is used later to study the flow fields local to a spherical inclusion in a quasi-infinite domain.

Fracture Plotting

The use of lattice elements simplifies fracture modeling in that only two nodes are involved in the local process of material separation, both in 2D and 3D, whereas multiple nodes are involved in continuum finite element models. However, the visualization of three-dimensional fracture processes is challenging for both types of analysis. In the authors' work, fracture is visualized by plotting the Voronoi facet of each fractured element or, to show crack openings, by plotting the Voronoi cells in the displaced configuration. Although the facet orientation differs from the crack band orientation when $\Theta \neq 0$, the Voronoi facets form a piecewise continuous representation of the crack surface (i.e. no overlaps or gaps are present). For example, Fig. 6 shows a series of three lattice elements, which are assumed to have fractured, and the corresponding representation of the crack trajectory. Furthermore, the facet areas are indicative of the relative size of the elements, whereas element size is generally not apparent when viewing the line element depiction of the lattice structure.

RESTRAINED DRYING SHRINKAGE OF SLAB STRUCTURE

Model Setup

Application of the lattice model is demonstrated for a basic, yet important, case: one-sided drying of a slab-like structure. Figure 7a shows relative dimensions of the slab structure (with d = 20 mm) and its Voronoi discretization. The lightly shaded cells along the top surface are associated with external nodes used to model the convective boundary conditions. The slab dries

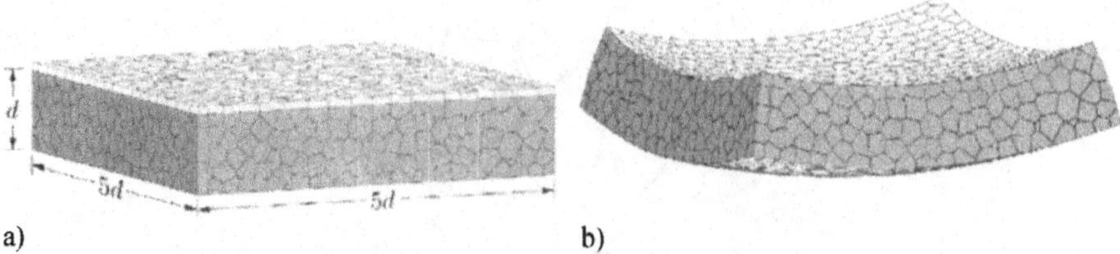

a) b)

Figure 7 a) Lattice model of slab structure and convective boundary conditions; and b) deformation due to drying shrinkage (no external restraint)

from the top surface and zero-flux boundary conditions are enforced elsewhere. Regarding the structural boundary conditions, two situations are considered: 1) the slab is unrestrained, except for at a single point to remove zero-energy modes of displacement; and 2) the nodes along the bottom surface are fully restrained, whereas the nodes along the vertical surfaces are free to move within the plane of the surface (to simulate periodic boundary conditions, at least prior to material fracture).

Coupled Moisture Diffusion – Fracture Analyses

The nonlinear diffusion and fracture parameters associated with Figs. 3 and 4b, respectively, have been assigned values typical of concrete materials, as determined through inverse analyses[6]. Figure 8 shows the cracking sequence up through a time t_s, where the cracking pattern has essentially stabilized (i.e. no new cracks form, although some cracks continue to open). Cracking forms what appear to be islands of material when viewed in plan. This general pattern is seen in the surface cracking of concrete and, more conspicuously, in the drying of clay-rich geomaterials (Fig. 9). The cracks tend to meet at triple junctures in both this field example and the numerical simulations. The average size of these surface patches appears to be a property of the system, but this issue is clouded by the small domain size and thus the unnatural boundary conditions that occur once fracture develops. However, the stabilization of the cracking pattern involves the unloading of some cracks (faintly visible in the transition from Fig. 8b to 8c, for example), which tends to indicate the size of the patches is a property of the system.

When viewing a cross-section of the cracked slab structure, it is clear that cracking branches laterally after advancing downward from the drying surface (Fig. 10). This type of branching has been seen in the drying shrinkage experiments of Bisschop[16] and in other numerical simulations of concrete materials[4]. This behavior is understandable in that each island of material continues to undergo drying shrinkage and therefore tends to curl upward, as does the unrestrained slab structure shown in Fig. 7b. Lateral cracking relieves stress in the surface region and thus acts to limit further fragmentation of the surface area.

The cracking process is driven by the moisture gradient that develops with drying. The degree to which the transport properties can affect the crack pattern is seen in Fig. 8d, which is produced using a constant average value of diffusivity (Fig. 3). The surface patches are smaller and, when viewed in 3D, lateral crack branching is limited. Future work will involve verifying the model results through comparisons with experimental data. To that end, the accurate determination of the input properties, including diffusivity, is important.

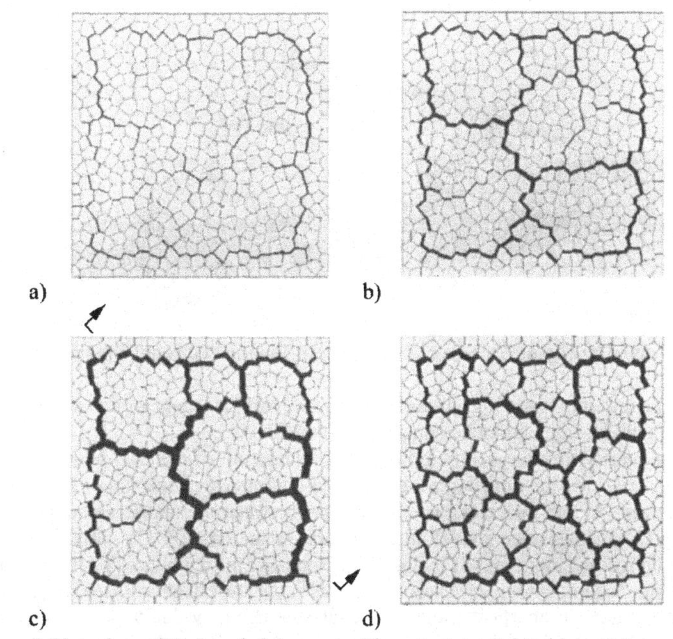

Figure 8 Plan view of drying shrinkage cracking at: a) $t = 0.4\ t_s$; b) $t = 0.6\ t_s$; and c) $t = t_s$ (nonlinear diffusion); and d) $t = t_s$ (constant average diffusion)

Figure 9 Cracking of clay-rich geomaterial due to drying

Crack patterns similar to those shown in Fig. 8 have been obtained from fracture analysis of thin solid films. Meakin[17] modeled a surface film as a 2D triangular lattice connected to a rigid substrate via elastic bond springs. For increasing tension in the surface film, fewer and longer cracks result from a more flexible bond to the substrate. More numerous cracks and diffuse cracking patterns result from a stiffer attachment to the substrate. The simulations presented in this paper differ in that fracture is driven by the gradient in pore humidity, which develops

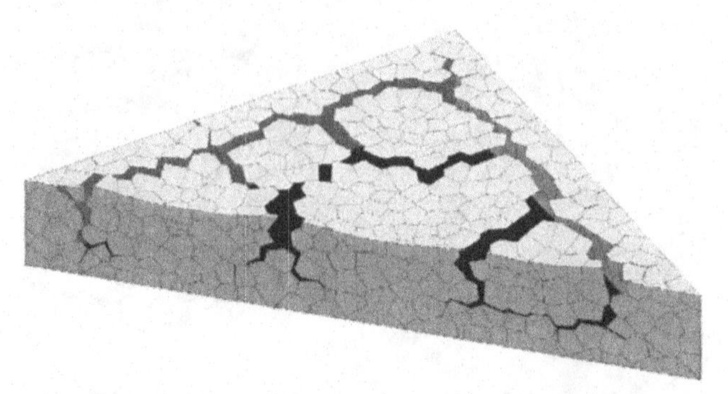

Figure 10 Sectional view of crack pattern (internal cells not shown)

through the slab depth due to drying. In addition, the flexural mechanism and associated lateral crack branching that develop cannot be simulated with a 2D lattice.

The preceding analyses have been based on the assumption of material homogeneity. As a next step, the concrete will be modeled as a three-phase composite consisting of a matrix phase, aggregate inclusions, and an interfacial transition zone between the matrix and aggregates. That approach is typical to the lattice modeling of concrete materials[18,19,20,21]. Within the coupled hygral-stress analyses, the modeling of inclusions will enable the study of: 1) the restraint of stiff inclusions and the accompanying effects on fracture[16]; 2) the effects of material structure on mass transport and other properties[22]; and 3) spatial heterogeneity due to material segregation, wall effects, etc. The following section presents preliminary work on the modeling of inclusions and their effects on the transport properties of the composite material.

ANALYSES OF MOISTURE TRANSPORT IN HETEROGENEOUS MEDIA

As a starting point for the mesoscale modeling of moisture transport in concrete materials, we look at steady-state incompressible, inviscid flow about a single spherical inclusion. This situation is governed by the Laplace equation, subject to appropriate boundary conditions. Results are provided for four cases: 1) homogeneous material (i.e. the inclusion has the same properties as the surrounding material); 2) an impermeable inclusion; 3) a spherical void; and 4) an impermeable inclusion surrounded by a porous ITZ.

A spherical inclusion of radius a is centrally located within a cubic material domain, as shown by the darker region in Fig. 11a. The sphere is discretized by pairs of nodes, as indicated by nodes i and j in Fig. 1c; each pair of nodes is equidistant from the sphere center and aligned in the radial direction. The nodal point density is greater close to the sphere to better resolve the flow field in that vicinity. Since the domain is large relative to the inclusion radius, comparisons can be made with theoretical solutions for flow about a sphere within an infinite domain.

Flow through a Homogeneous Material

If the inclusion is assigned the same transport properties as the surrounding material, the numerical solution should agree with the theoretical solution for a homogeneous medium. For a difference of pressure potential, P_2-P_1, acting over the domain in the x-direction (with zero-flux

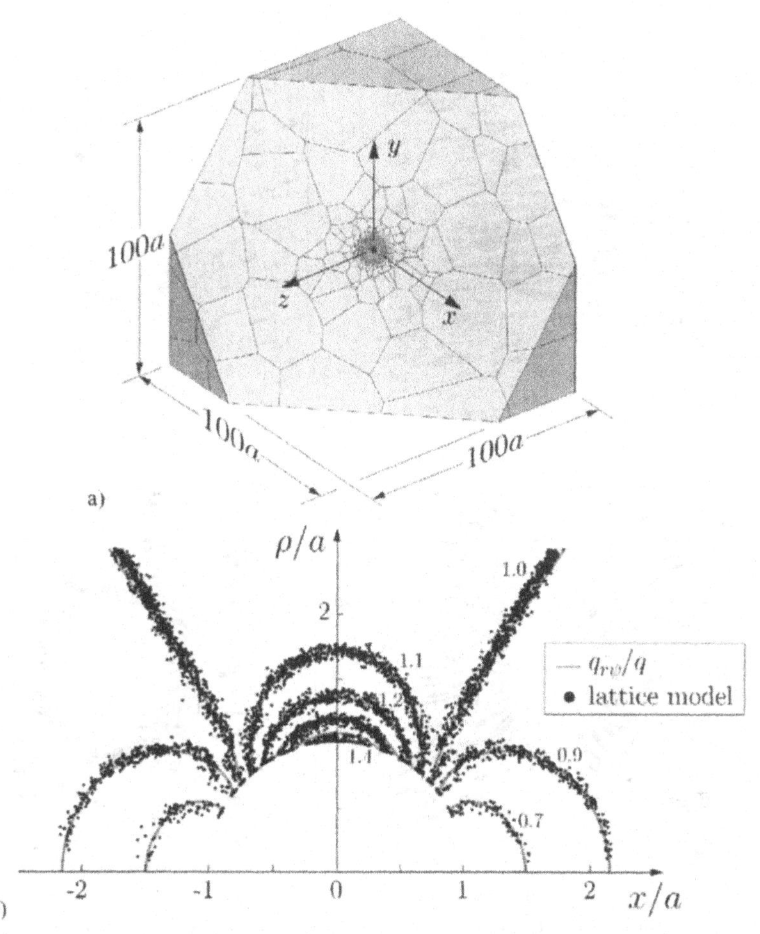

Figure 11 a) cross-section of spherical inclusion within quasi-infinite domain; and b) iso-flux contours local to sphere

boundary conditions on the *y*- and *z*-faces), the flow intensity is uniform throughout the domain and can be calculated using Darcy's Law

$$q = K\frac{(P_2 - P_1)}{L} \tag{11}$$

where *K* is the permeability coefficient and *L* is the length of the flow path (= 100*a*). With no loss of generality, it is assumed that $K = 1$. The computed flux field is shown in Fig. 12a for a thin slice of the material (i.e. for $-0.2 < z/a < 0.2$). For all plots given in Fig. 12, the flux vectors

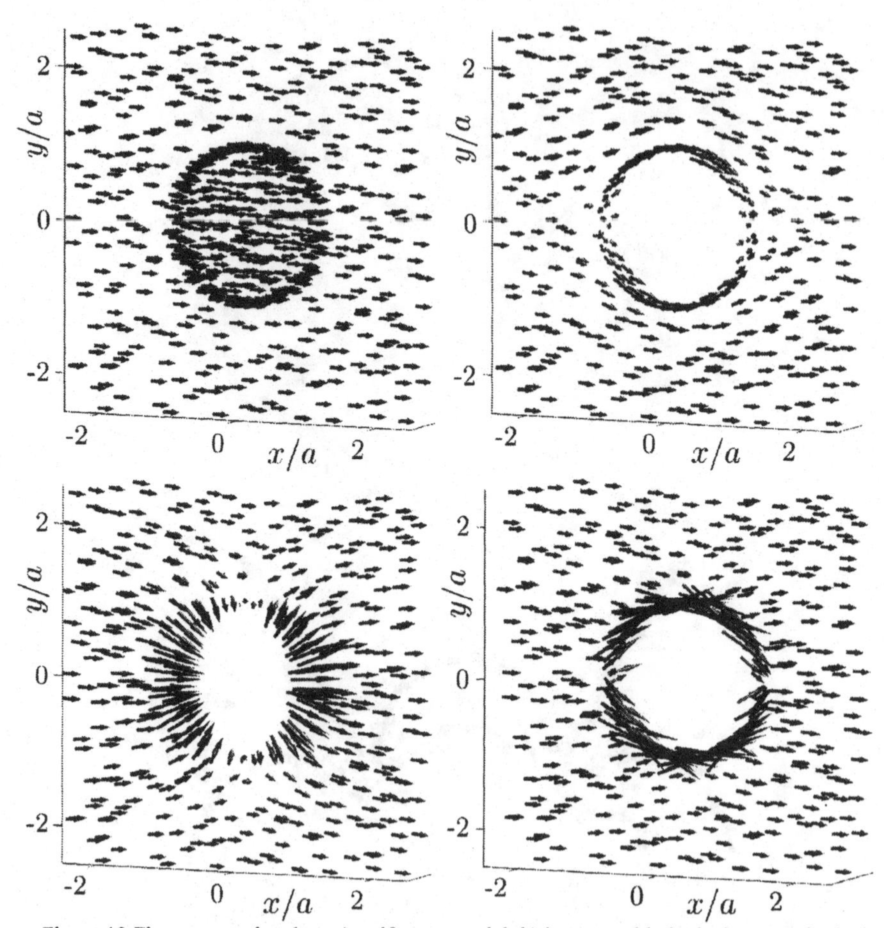

Figure 12 Flow vectors local to: a) uniform material; b) impermeable inclusion; c) spherical void; and d) impermeable inclusion with highly permeable ITZ

are determined by the mass conservation scheme depicted in Fig. 5 (where the spherical coordinates θ and ϕ are chosen to maximize the flux magnitude.) The large number of flux vectors within the inclusion region is due to the abundance of nodes in that region. Over the entire domain, the nodal flux values have an average value of $1.00000026q$ with a standard deviation of $1.16\text{x}10^{-5}q$. It is clear that mesh geometry does not affect the results for this uniform flow field. The close agreement with theory is due to the A_{ij}/h_{ij} scaling of the elemental relations (Eq. 5), as determined from the Voronoi diagram[5,6].

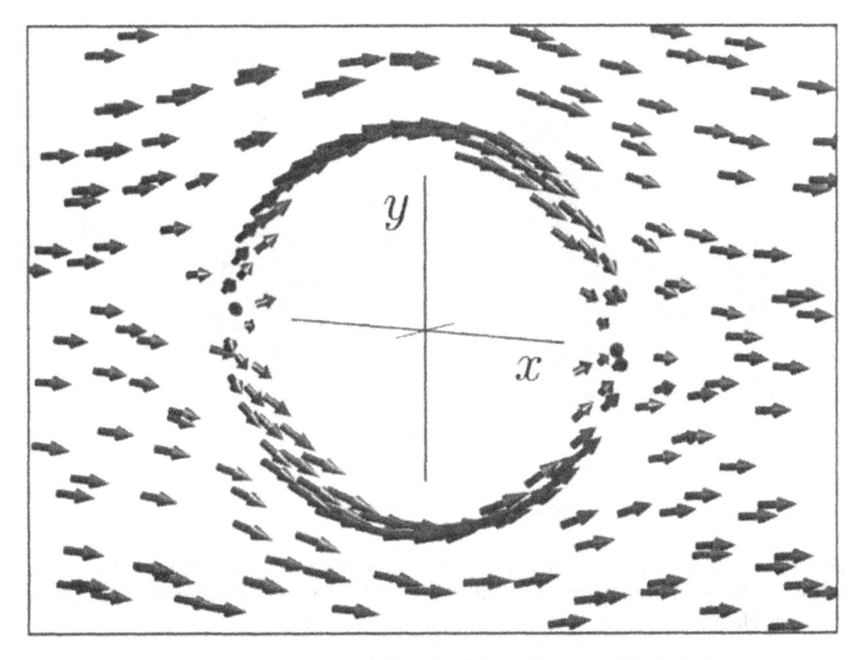

Figure 13 Detailed view of flow field about impermeable inclusion

Flow Local to an Impermeable Inclusion

For the case where the spherical inclusion is impermeable, flow is diverted around the inclusion as shown in Fig. 12b and, in greater detail, in Fig. 13. The accuracy of these results is shown by the iso-flux points presented in Fig. 11b, which are produced by interpolating between nodal values. Since the flow field is axisymmetric about the x-axis, all possible results have been plotted in the x-ρ plane, where $\rho = (y^2+z^2)^{0.5}$. The computed flux values agree well with the corresponding theoretical solution[23]

$$q_{r\psi} = \sqrt{q_r^{\,2} + q_\psi^{\,2}} \tag{12}$$

in terms of the flux components in the r and ψ directions. These are given by

$$q_r = q\cos\psi\left(1 - \left(\frac{a}{r}\right)^3\right) \tag{13}$$

$$q_\psi = -q\sin\psi\left(1 + \frac{1}{2}\left(\frac{a}{r}\right)^3\right) \tag{14}$$

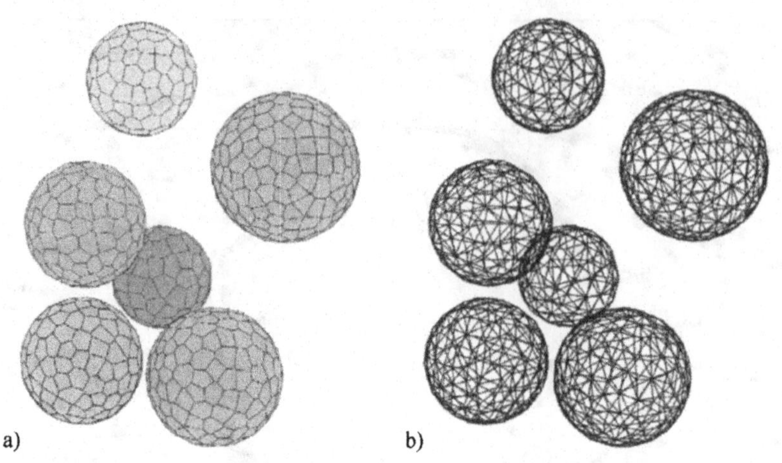

Figure 14 a) Voronoi discretization of spherical inclusions; and b) triangulation of ITZ layer for modeling of moisture transport

where $r = (x^2+y^2+z^2)^{0.5}$ and ψ is the angle r makes with the x axis. The scatter in the numerical values is due, in part, to the relatively coarse discretization near the inclusion using these low-order lattice elements.

Flow Local to a Spherical Void

The spherical void is modeled by assigning large K values to the lattice elements within the sphere. For this case, flow is drawn inward to the void as shown in Fig. 12c. For the sake of clarity, the large flow vectors for nodes within the sphere have not been plotted. In contrast with the previous case, flow along the x-axis is magnified rather than impeded. For both cases, the flow field is anti-symmetric about the $x = 0$ plane.

Flow Local to an Impermeable Inclusion with a Porous ITZ

Lattice elements spanning the ITZ, as shown in Fig. 1c, are most relevant when modeling fracture of the ITZ. For modeling moisture transport, however, the properties of the lattice parallel to the ITZ are also important. As part of the Delaunay tessellation of the domain, the pairs of nodes used to discretize each sphere are triangulated to form two concentric lattice structures. As an example, triangulation of the outer layer of nodes is shown in Fig. 14 for several spheres. For the previous simulations, the inner layer represented the inclusion (or void), whereas this outer layer represented the matrix phase.

In this section, the outer layer approximates the ITZ between the inclusion and matrix phases. Whereas this approach provides a well-defined boundary between the ITZ and the inclusion, the distinction between the ITZ and the surrounding matrix material is not clear. Rather, the A_{ij} of the outer layer of elements are strongly affected by the neighboring random distribution of nodes representing the matrix. For the simulation given in this section, the average thickness of this layer of elements (in the sphere radial direction) is significantly larger than the target ITZ thickness. For this reason, the permeability properties of these elements are based on a weighted

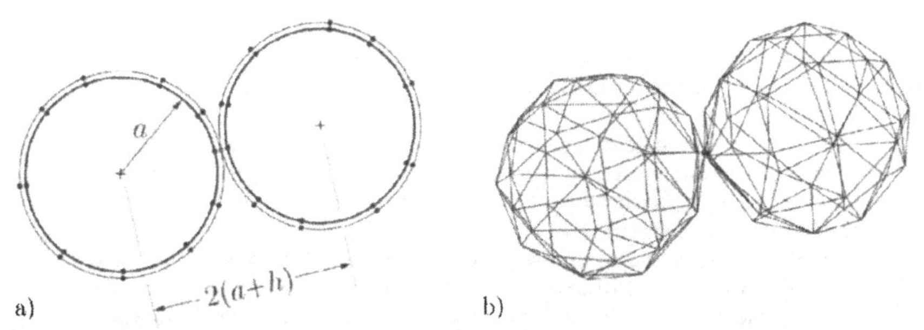

Figure 15 Establishing local connectivity of flow paths: a) 2D schematic representation; and b) 3D inclusions

(mixture rule) average of the desired ITZ and matrix permeabilities. This uncertainty in the modeling of ITZ properties can be avoided by introducing a third node to each pair of nodes. The middle layer of nodes, representing the ITZ, would then be bounded on each side by two regular structures, allowing for precise control of the ITZ thickness.

Figure 12d shows a slice of the flow field about an impermeable inclusion with a permeable ITZ. The ITZ is assumed to have a thickness of $a/10$ (which is rather large for sizeable a) and a relative permeability of $K_{ITZ} = 100K_m$, where K_m is the permeability of the matrix. As noted above, the K value assigned to the outer layer of lattice elements is determined using an ordinary mixture rule, which resulted in $K << K_{ITZ}$. Although the modeling of the ITZ is still rather crude, it is clear that ITZ acts as a channel for moisture transport to the extent that the surrounding flow field is significantly impacted. Similar to the case of the spherical void, the highly permeable ITZ draws flow toward the inclusion and the ability of the inclusion to impede flow is not as evident, as it is Fig. 12b. A range of behaviors can be simulated, depending on the relative properties of the different phases.

Our future efforts will involve improvement of the ITZ model and more quantitative analyses of the flow properties. In addition, the transport properties of the ITZ can be modified to reflect cracking along the interface, as determined from the fracture analyses. One primary interest is in the modeling of percolation phenomena through the interconnection of ITZ elements. Figure 15 shows a meshing strategy for establishing a single node connection in both two and three dimensions. The meshing of ITZ overlap through multi-node connections is more challenging, as is modeling an appropriate size distribution of aggregate inclusions.

CONCLUSIONS

Shrinkage cracking exposes the internal concrete and reinforcement to the surrounding environment and therefore is a contributing factor to various durability problems. The potential for such cracking can be reduced through: 1) proper mix design, including the use of special admixtures; and 2) avoiding unfavorable forms of restraint against free shrinkage. The situation is complicated in that restraint is generally present in several forms and to varying degrees, depending on the interrelationships between the material parameters, microclimate, structural dimensions, and structural boundary conditions. Computer modeling is a means for studying these interrelationships and estimating the potential for shrinkage cracking.

In this paper, an irregular lattice model is used to simulate drying shrinkage and cracking of cement composites. Macroscopic models of moisture diffusivity and fracture are both defined for a statistically homogeneous material. This paper also reports on basic simulations of moisture transport at the mesoscale, where the material matrix, inclusions, and the ITZ are modeled as discrete entities within the lattice model. Several conclusions can be made.

1. Even for the basic case of one-sided drying of a slab structure, the simulated cracking patterns indicate that cracking is driven by complex, three-dimensional processes. For the simulations of moisture transport at the mesoscale, as well, the need for three-dimensional models is apparent.
2. The modeling of moisture diffusivity can affect the crack patterns that develop in slab structures. Changes in diffusivity affect the humidity gradient through the slab thickness and thus the stress profile. For the cases considered here, lateral branching of cracks occurred for a nonlinear dependence of diffusivity on relative humidity, whereas such cracks did not occur for a constant average diffusivity value. The results suggest that the lateral branching of cracks reduces the degree of restraint. Thus, the potential for surface cracking is also reduced (i.e. the lateral branching of cracks tends to increase average crack spacing on the surface).
3. A method is given for calculating nodal flux, which is based on the conservation of mass transported through a node by the adjoining lattice elements. This method is useful for visualizing the flow fields and for verifying the accuracy of the lattice model. Nodal flux values given by the model agree precisely with theory for the basic case of uniform flow through a homogeneous medium. These results are independent of the irregular geometry of the lattice. Good agreement with theory is obtained for flow about an impermeable spherical inclusion in an infinite domain.
4. The Delaunay-Voronoi dual construction of the material domain provides an intuitive means for explicitly modeling the ITZ between inclusions and the matrix phase, at least for inclusions with simple geometry. Abilities of the lattice to model flow pathways local to inclusions, as affected by the properties of the ITZ, were demonstrated through several simulation examples.
5. The basic results presented in this paper indicate a potential for extending this lattice approach to model more general materials and structures. Future efforts will be directed toward simulating drying shrinkage cracking within the concrete mesostructure, and the modification of transport properties as a result of such cracking.

REFERENCES

[1] Comite Euro-International du Beton (CEB), "Durable Concrete Structures - Design Guide," Thomas Telford Ltd., London, 1992.

[2] ACI Committee 446, "Fracture mechanics of concrete: Concepts, models and determination of material properties," in: Z.P. Bazant (Ed.) Fracture Mechanics of Concrete Structures, Elsevier Applied Science, London, 1-140, 1992.

[3] H. Sadouki, J.G.M. van Mier, "Analysis of hygral induced crack growth in multiphase materials," *HERON*, **41**(4), 267-86 (1996).

[4] J.E. Bolander, S. Berton, "Simulation of shrinkage induced cracking in cement composite overlays," *Cement and Concrete Composites*, **26**, 861-71 (2004).

[5] J.E. Bolander, N. Sukumar, "Irregular lattice model for quasistatic crack propagation," *Physical Review B*, **71**, 094106 (2005) (12 pages).

[6]S. Berton, J.E. Bolander, "Crack band modeling of fracture in irregular lattices," *J. Computer Methods in Applied Mechanics and Engineering*, (2005) (In press).

[7]Z.P. Bažant, L.J. Najjar, "Drying of concrete as a nonlinear diffusion problem," *Cement Concrete Research*, **1**, 461-73 (1971).

[8]G. Martinola, F.H. Wittmann, "Application of fracture mechanics to optimize repair mortar systems," in: F.H. Wittmann (Ed.), Fracture Mechanics of Concrete Structures, AEDIFICATIO Publishers, Freiburg, 1481-86, 1995.

[9]J. Marchand, B. Gérard, "New developments in the modeling of mass transport processes in cement-based composites -- a review," in: Advances in Concrete Technology - Proceedings Second International CANMET/ACI International Symposium (SP-154), ACI, 169-210, 1995.

[10]Z. Li, Y.M. Lim, J.E. Bolander, "Role of fibers during non-uniform drying of cement composites," in: V.C. Li et al. (Eds.), International RILEM Workshop on HPFRCC in Structural Applications, Honolulu, May, 2005.

[11]A. Okabe, B. Boots, K. Sugihara, "Spatial Tessellations: Concepts and Applications of Voronoi Diagrams," Wiley Series in Probability and Mathematical Statistics, John Wiley & Sons Ltd., England, 1992.

[12]R.W. Lewis, K. Morgan, H.R. Thomas, K.N. Seetharamu, "The Finite Element Method in Heat Transfer Analysis," John Wiley & Sons, Chichester, UK, 1996.

[13]T. Kawai, "New discrete models and their application to seismic response analysis of structures," *Nuclear Engineering and Design*, **48**, 207-29 (1978).

[14]Z.P. Bazant, B.H. Oh, "Crack band theory for fracture of concrete," *Materials and Structures*, RILEM, **16**(93), 155-77 (1983).

[15]G. Strang, "Introduction to Applied Mathematics," Wellesley-Cambridge Press, Wellesley, MA (1986).

[16]J. Bisschop, "Drying shrinkage microcracking in cement-based materials," Ph.D. Thesis, Delft University of Technology, Delft, 2002.

[17]P. Meakin, "Simple kinetic models for material failure and deformation," in: H.J. Herrmann and S. Roux (Eds.), Statistical Models for the Fracture of Disordered Media, Elsevier Science Publishers, North-Holland, 291-320, (1990).

[18]E. Schlangen, J.G.M. van Mier, "Experimental and numerical analysis of micromechanisms of fracture of cement-based composites," *Cement and Concrete Composites*, **14**, 105-18 (1992).

[19]E. Schlangen, E.J. Garboczi, "Fracture simulations of concrete using lattice models: computational aspects," *Engineering Fracture Mechanics*, **57**, 319-32 (2/3) (1997).

[20]G. Lilliu, J.G.M. van Mier, "3-D lattice type fracture model for concrete," *Engineering Fracture Mechanics*, **70**, 927-41 (2003).

[21]G. Cusatis, Z.P. Bažant, L. Cedolin, "Confinement-shear lattice model for concrete damage in tension and compression: I. Theory," *J. Engineering Mechanics*, **129**(12), 1439-58 (2003).

[22]D.P. Bentz, E.J. Garboczi, K.A. Snyder, "A hard core/soft shell microstructural model for studying percolation and transport in three-dimensional composite media," NISTIR 6265, National Institute of Standards and Technology, U.S. Department of Commerce, 1999.

[23]R.H. Kirchhoff, "Potential Flows – Computer Graphics Solutions," Dekker, New York, 1985.

A SIMULATION MODEL OF THE PACKING ARRANGEMENTS OF CONCRETE AGGREGATES

Konstantin Sobolev* and Adil Amirjanov**

*Facultad de Ingeniería Civil
Universidad Autónoma de Nuevo León
San Nicolas, NL, México

** Department of Computer Engineering
Near East University
Nicosia, N. Cyprus

ABSTRACT

The behavior of particulate composite materials, such as portland cement concrete, depend to a large extent on the properties of their main constituent – the aggregates. Among the most important parameters affecting the performance of concrete are the packing density and corresponding particle size distribution of aggregates. Better packing of aggregates improves the main engineering properties of composite materials: strength, modulus of elasticity, creep, shrinkage. Further, it brings major savings due to a reduction in the volume of binder.

A simulation algorithm was developed for the modeling of packing of large assemblies of particulate materials (in the order of millions). These assemblies can represent the real aggregate systems composing portland cement concrete. The implementation of the developed algorithm allows the generation and visualization of the densest possible arrangements of aggregates. The influence of geometrical parameters and model variables on the degree of packing and the corresponding distribution of particles was analyzed. Based on the simulation results, different particle size distributions of particulate materials are correlated to their packing degree. The obtained aggregate packings show a good agreement with the standard requirements and available research data. The developed models can be applied to the optimal proportioning of concrete mixtures.

INTRODUCTION

The properties and behavior of particulate composite materials, such as portland cement concrete depend, to the large extent, on the properties of their main constituent – aggregates. It is commonly accepted that particle size distribution (PSD) is among the most important characteristics of the aggregates [1, 2].

The first attempts to provide the "best" optimal particle size distribution were based on trials with balls of different diameters [3-7]. These experiments resulted in optimal distribution curves which are currently accepted as standards [6]. One of the early examples presented by Fuller [8] is a series of curves which are widely used for the optimization of concrete and asphalt aggregates:

$$P_i = 100 \left(\frac{d_i}{D_{max}} \right)^n$$

where:

- P_i total percent of particle passing through (or finer than) sieve;
- D_{max} maximal size of aggregate;
- d_i diameter of the current sieve; and
- n exponent of the equation (0.45-0.7).

Because it is relatively simple to achieve the "target" distribution with a minimum deviation using a few (at least two) sets of particulate materials [9, 10], this optimal distribution method is extensively used. Yet, in spite of its usefulness, this method cannot predict the packing degree of the resultant mixture.

With the development of computers the packing problems of real systems became a challenging subject for engineers. The first computer algorithms were able to pack less than 1000 spheres per hour [4]. Such algorithms are based on the modeling of the movement of a particle (usually represented by a sphere or ellipsoid) due to rolling or sliding under the compaction gradient. Based on this strategy, the particles in a rigid container are forced to occupy the best vacant positions within the neighborhood. Modern modeling approaches include better insight into the natural packing process or full-scale modeling of the experiment [11-29]. The contribution of additional factors (such as friction and deformation) acting at the contact points was found to be essential for the modeling of the dynamic processes, involving the particulate materials [13-19]. Usually each movement of a particle requires the solution of the differential equation; this procedure slows down the calculation process. A comprehensive review of packing algorithms was provided elsewhere [13-17, 27].

It is believed that optimal particle size distribution (PSD) corresponds to the "best" or the densest packing of the constituent particles. However, the modeling of packing of large particulate assemblies had demonstrated that the densest arrangements (PSD) of particles are actually not realized in concrete technology [5, 7, 10, 30]. As it was shown, only "gap-gradings" could be considered to some extent as a sort of dense arrangement of particles. The majority of "practical" concrete aggregates gradings lying between the 0.45-0.7 power curves are actually "loose - initially - packed", LIP systems [5, 7, 10, 30]. The fundamental characteristic of such systems is related to a wide range of particle sizes required to achieve the high packing degree at limited number of largely-sized particles. These are opposed to "perfect" geometric, regular arrangements with high a density that is realized within a relatively narrow size range [5, 7].

IMPORTANCE OF THE PROBLEM

In spite of recent progress in the development of packing algorithms, it is evident that a new approach is needed to model the packing of large assemblies of particulate materials representing the aggregate structure of portland cement concrete. On the one hand, the natural packing processes of large particulate assemblies must be imitated; on the other hand, the approach must be easily applicable to solve practical problems. Of particular interest for practical application are the problems of reconstructing the particulate structure represented by concrete aggregates and establishing the relationship between the particle size distribution and the degree of packing.

DESCRIPTION OF THE PACKING MODEL

It was found that a good approximation of particulate systems of elementary volume within a container with a rigid or periodic boundary can be achieved when the center of the particle is randomly located at the grid of a cubic lattice [5, 7, 10, 30]. In this case, a thick 3D mesh with an opening size of less than 1/100 of the minimal diameter of the particle must be used to minimize any possible error. The particle is considered as a discrete element which is represented by a sphere.

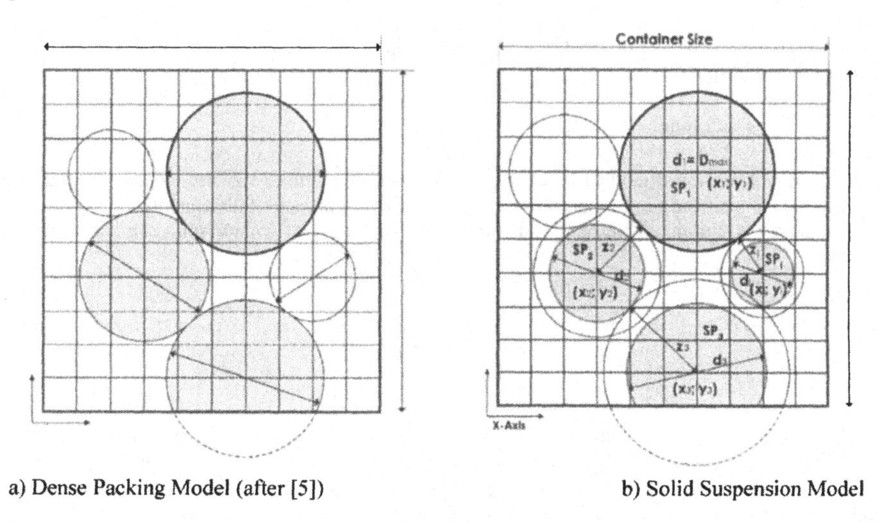

a) Dense Packing Model (after [5]) b) Solid Suspension Model

Fig. 1. 2D Representation of the Packing Models

Sequential Packing Model

The simulation takes place in a cube C(1), where 1 is a length of side. A two-dimensional representation of sequential packing model of spheres is shown in the Fig. 1. New spherical particles are sequentially placed into the cube with the center glued to the node of a lattice grid (with a lattice grid size of 1/32766) and with radius in the range $r_{min} < r \leq r_{max}$. The radius r_{max} is fixed experimentally but r_{min} is decreased gradually so allowing larger spheres to be placed in a cube prior the placement of smaller ones. The formula for calculating r_{min} is as follows:

$$r_{\min(n)} = \frac{r_{\max}}{(k_{red})^n} \tag{1}$$

where $k_{red} = 1+10^K$ is a constant for reduction the r_{min}, K is the reduction coefficient and n is a number of the packing attempts (steps) required to gradually reduce r_{min}.

Initially, the cube is prepacked with an initial sphere (or spheres; Fig. 1a shows the prepacked sphere SP1). Then the center of a new sphere is generated randomly within the cube lattice. Before locating the sphere with radius r the various conditions are checked:

- the center of the new sphere can not be located inside of any already packed spheres;
- the new sphere can not cross any already packed spheres (no overlapping of spheres is permitted);
- the minimum distance to the surface of any already packed spheres should be greater than r_{min}.

Fig. 1a shows the spheres SP_2, SP_3 and SP_i are placed in the case if the d_2, d_3 and d_i, respectively satisfy the conditions above.

Solid Suspension Model

The Solid Suspension Model assumes that the particles of a similar size are located at a maximal possible distance from each other, imitating perfect mixing process. Subsequently, all spheres are separated by a certain distance δ and, therefore, the radius of a new sphere is calculated as follows:

$$r = \frac{z}{(1 + \frac{k_{del}}{s^m})} \tag{2}$$

where z is a minimum distance to the surface of any already packed spheres, k_{del} is an initial coefficient to provide the separation between spheres, $s = 1+10^{Sr}$ is a constant to reduce the separation distance, S_r is the separation reduction coefficient and m is the number of the packing attempts (steps) required to gradually reduce this coefficient.

In the Fig. 1b the spheres SP_2, SP_3 and SP_i are located with separation; and the corresponding radii are calculated according to the expression (2). This arrangement allows particles of similar size be located at the specified maximal possible distance from each other and avoid the formation of congested zones or conglomerates; this also would imitate the "perfect" mixing process. Finally, the radius of any new sphere should satisfy the following conditions that are calculated by combining the expressions (1) and (2):

$$\begin{cases} r \geq \dfrac{r_{max}}{(k_{red})^n * (1 + \dfrac{k_{del}}{s^m})} \\ r \leq r_{max} \end{cases} \tag{3}$$

If the sphere does not meet the requirements (3), this sphere is discarded and the center of a new sphere is randomly generated. This process continues until the amount of packing attempts (N) is spent. At this point, the variables m and n are incremented by 1 and the process of placing the spheres is continued for the spheres of smaller radii. However, the incrementing procedure is limited by the following conditions:

$$(k_{red})^n \leq 256 \text{ and } \frac{k_{del}}{s^m} \leq 10^{-6}$$

The packing of a cube will be terminated as all initially requested spheres are packed or a number of steps required to reduce r_{min}, n has reached a limit. By the end of the packing process of each sphere the volume fraction of solid particles is calculated:

$$V = \sum_{i=1}^{N} V_i$$

where V_i is a volume of a particular sphere and N is the number of the spheres packed. Based on the individual volumes of newly packed spheres, the particle size distribution of particles and packing degree are updated.

RESEARCH PROGRAM

In this research program, the packing into a container with periodic boundaries was considered in order to represent the elementary volume of particulate composite and to eliminate the wall effect. The ratio of the container size to the maximum diameter of the sphere was fixed at 3.3 (that is a common assumption related to the density measurements when the wall effect is eliminated). The total amount of spheres used in the packing trials (N_{total}) varied from 1 to 10 millions (1M for the experiment A and 1M-10M for the experiment B) and the mesh size (pixel) was 1/32766 of the container length.

The research program considered the investigation of only solid suspension model, i.e. the separation between the particles was always set. Based on the preliminary evaluation, the separation reduction coefficient was set to -3 providing a reasonably slow reduction of the separation distance. The reduction coefficient was varied from -1 to -3; at higher values providing a somewhat quicker drop in the minimal size of the particles and also resulting in the formation of loose or "diluted" packing arrangements. The number of packing trials per cycle, N was equal to 10^5, the same value for both particle size reduction and separation reduction.

The variable parameters and their levels for the experiments A and B are presented in Table 1.

RESULTS AND DISCUSSION

The results of the simulation algorithm are presented in Figs. 2 - 8, where Figs. 2, 4, 6 demonstrate packing progress and generated PSD with the passing values given for a specific particle sizes (that are standard for the sieve analysis and determined by a formula: $d_i = D_{max}/2^m$, where D_{max} is the maximal size of the sphere and m = 0, 1, ... , k). Figs. 3, 5, 7 provide the visualization of the packing patterns obtained with 1000 spheres at a reduction coefficient K = -3.

Table 1. The Parameters of the Simulation Model

Experiment	Total Amount of Spheres, N_{total}	Number of Packing Trials, N	Reduction Coefficient, K	Separation Coefficient, K_{del}	Separation Reduction Coefficient, S
A	$1M = 10^6$	$100k = 10^5$	-3	0.1	-3
			-2	0.5	
				1	
			-1	1.5	
B	$1M = 10^6$	$100k = 10^5$	-3	1.5	-3
	$2M = 2*10^6$				
	$5M = 5*10^6$				
	$10M = 10^7$				

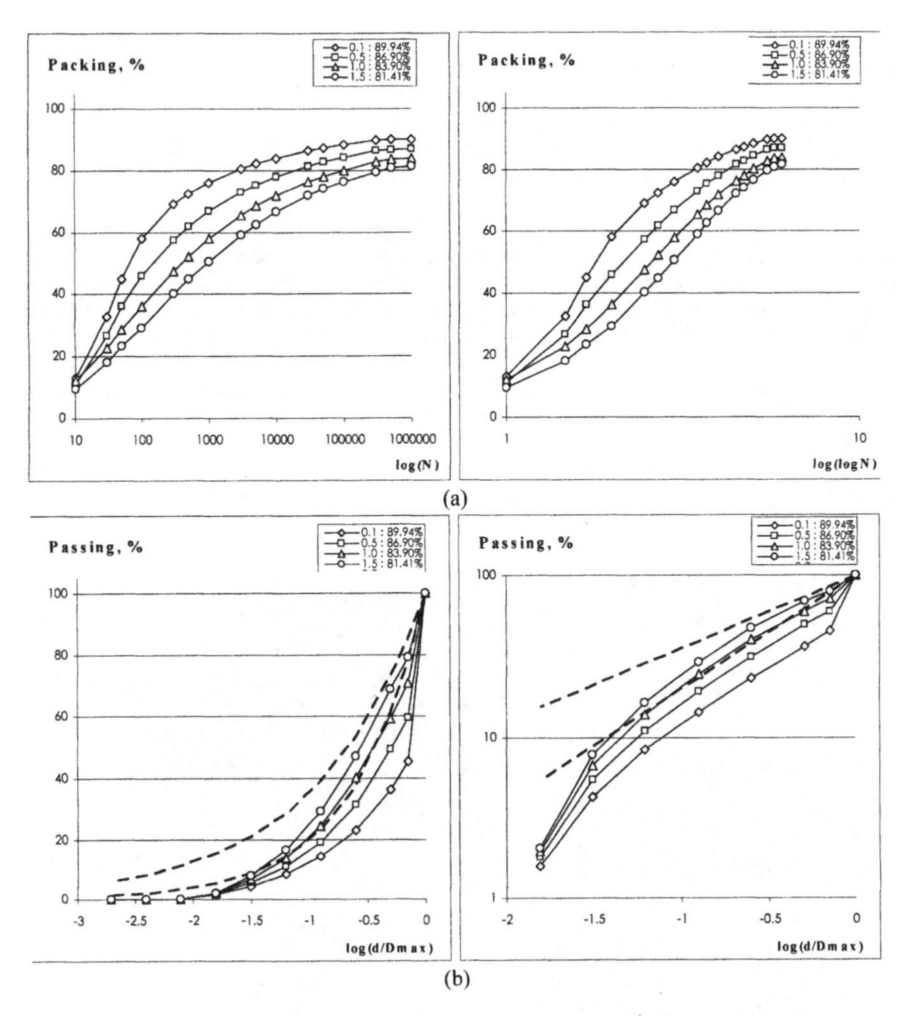

Fig. 2. The application of the packing model for packing of 10^6 spheres at a reduction coefficient -3: (a) development of packing degree and (b) particle size distribution

Packing Process

The development of the packing degree for different values of the reduction coefficient, K and separation coefficient, k_{del} is demonstrated in Figs. 2a, 4a, 6a. It can be observed that the major contribution to packing is provided by the first 10^5 (100k) packed spheres. It is important that the packing of the first 100 spheres is represented by a straight line which is determined by the combination of the reduction and separation coefficients. The combination of low reduction coefficient (K = -3) and low separation coefficient (k_{del} = 0.1) results in the densest possible packing. There is no significant density increase after the packing of 100k spheres in this case, resulting in the highest value of packing degree of 89.94%.

Increase of separation coefficient to its maximal value provides the generation of a distribution of particles with significant portion of mid-size particles (at a predominant size of $D_{max}/2$ - $D_{max}/8$). This results in a slow increase of packing densities vs. number of particles. However, at a low reduction coefficient (K = -3) the increase in separation coefficient results in a high value of packing degree (81.41%) and such arrangement is considered as densely packed.

The increase in the reduction coefficient provides an increase of the deviation between the packing curves and the considerable reduction in packing degree. With a high value of reduction coefficient (K = -1) the increase of separation coefficient (from 0.1 to 1.5) results in considerable reduction of the packing degree from 81.77 to 38.70%. This is mainly due to relatively large distances set between the particles and corresponding surplus of intermediate phase.

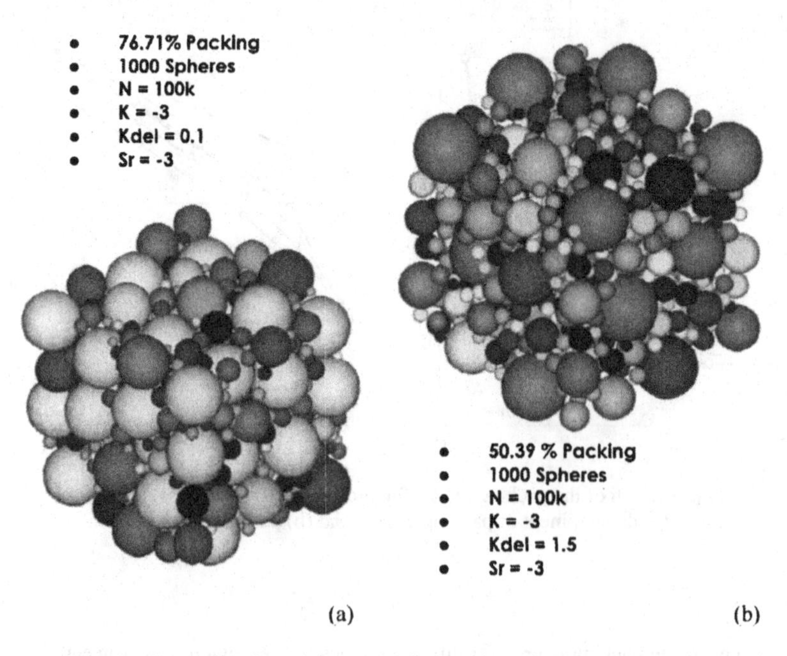

Fig. 3. The application of the model for packing of 1000 spheres at a reduction coefficient -3 and separation coefficient 0.1 (a) and 1.5 (b)

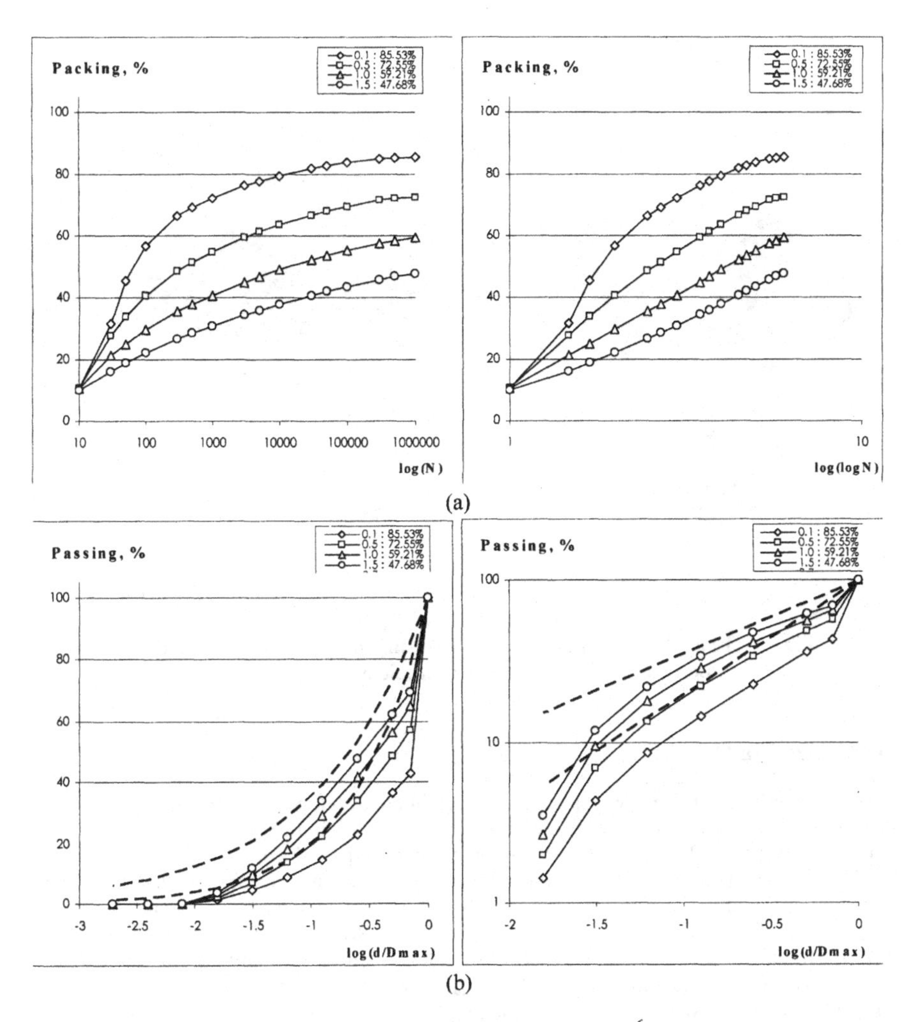

Fig. 4. The application of the packing model for packing of 10^6 spheres at a reduction coefficient -2: (a) development of packing degree and (b) particle size distribution

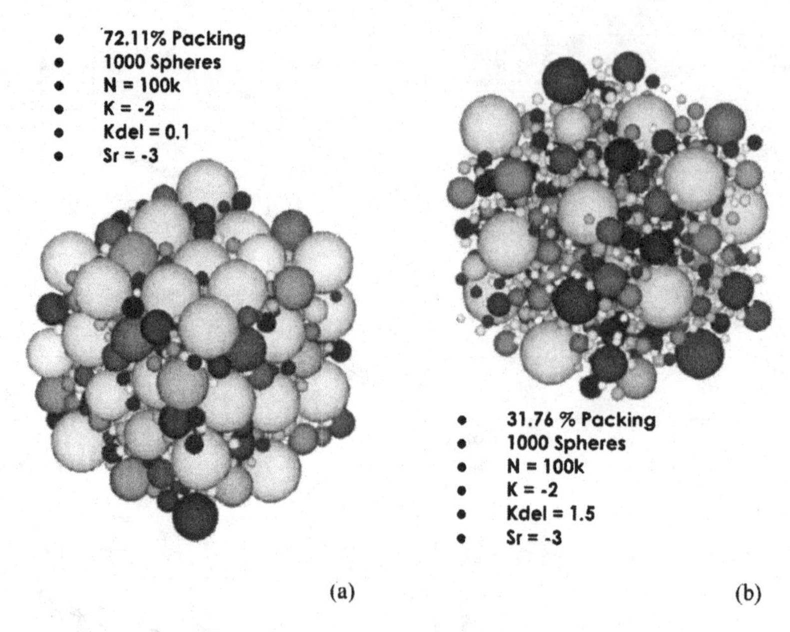

Fig. 5. The application of the model for packing of 1000 spheres at a reduction coefficient -2 and separation coefficient 0.1 (a) and 1.5 (b)

Particle Size Distribution

The major interest of the model application is seen in the development of the distribution curves that match those used in the concrete technology (Figs. 2b, 4b, 6b). The best (the most dense) gradings with 81 - 90% packing degree are obtained with the reduction coefficient at its lowest level -3 (at all separation coefficient values) or with the separation coefficient 0.1 (at all values of reduction coefficient). However, it can be observed that the curves, obtained with low separation coefficient (i.e. k_{del} = 0.1) representing the densest assemblies, are "modified" Fuller type or "Initially Pre-Packed" gradings (IPP - gradings) with a predominant volume (~50%) of largest particles ranging from D_{max} to 0.85*D_{max} [5, 7]. Here, the arrangement of this group of particles results in about 50% of the packing; and about 30% of the sphere's volume is represented solely by the spheres of maximal size (D_{max}) which account for 25% of the packing. Finally, a relatively narrow range of particle sizes (from D_{max} to $D_{max}/2$) provides about 60% of the packing (Fig. 3a, 5a, 7a). [5, 7]. This group of particles is considered to be arranged in a manner similar to the "ideal" regular close-packed lattices (Fig. 3a) approaching the condition of the maximum possible value for randomly packed systems (jammed state) [5, 7, 21]. Such dense packing conditions are determined by a low separation coefficient at different values of reduction coefficient. In this case, the distribution of particles is represented by the relatively narrow range of the sizes. These distributions are characteristic of only "gap-graded" aggregate mixtures which lay outside the limits commonly used in concrete technology (i.e. outside the limits set by

Fuller curves with the exponent from 0.45 to 0.7). To achieve the same packing degree at a less arranged initial structure (LIP condition) a much wider range of sizes is necessary (Fig. 2b). The graphical representation of this packing arrangement is given in Fig. 3b.

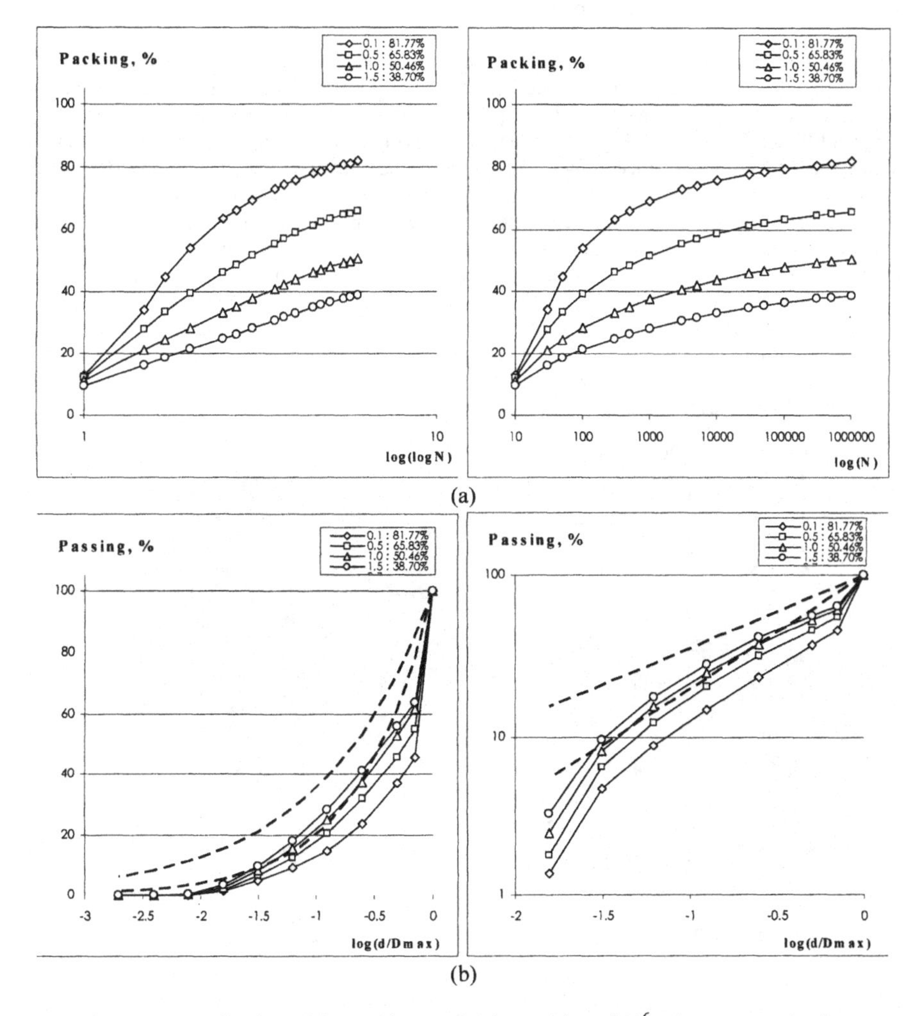

Fig. 6. The application of the packing model for packing of 10^6 spheres at a reduction coefficient -1: (a) development of packing degree and (b) particle size distribution

However, in practice, the achievement of well-arranged initially pre-packed structures is quite difficult using conventional compaction methods due to friction between particles and their irregularity. Therefore, many "real" particulate assemblies could be described by the models with low separation coefficient. This condition is the case of "Loose Initial Packing" (LIP) arrangement, when the largest particles (from D_{max} to $0.85*D_{max}$) occupy less than 25% of the volume and provide less than 25% of the packing [5, 7].

The LIP condition can be realized with the solid suspension model at a low reduction coefficient (-3) and high value of separation coefficient (0.5). The corresponding particle size distribution curve lies between the conventionally employed levels of the Fuller distribution curve with exponent 0.45 and up to 0.7 (Fig. 2b, 3b, 8). The packing density of this particulate assembly is higher than the common concentration of concrete aggregates, i.e. 70%; this implies that some additional separation would take place in a real system. Such a condition might be obtained by an increase in the separation coefficient beyond its maximal level. Further an increase of the reduction coefficient, up to -1 at a high value of separation coefficient (0.5 and more) leads to relative increase of coarsely sized fractions within the mix, but results in a low concentration of solids. The packing degree or, actually, concentration of spheres corresponding to an arrangement with a reduction coefficient of -1 and separation coefficient 1.5 is only 38.7% and corresponding particle size distribution is characterized by about 40% of large particles (in the range from D_{max} to 0.85 D_{max}) or IPP condition. Such systems may describe the arrangement of particles in diluted suspensions (Figs. 5b, 7b).

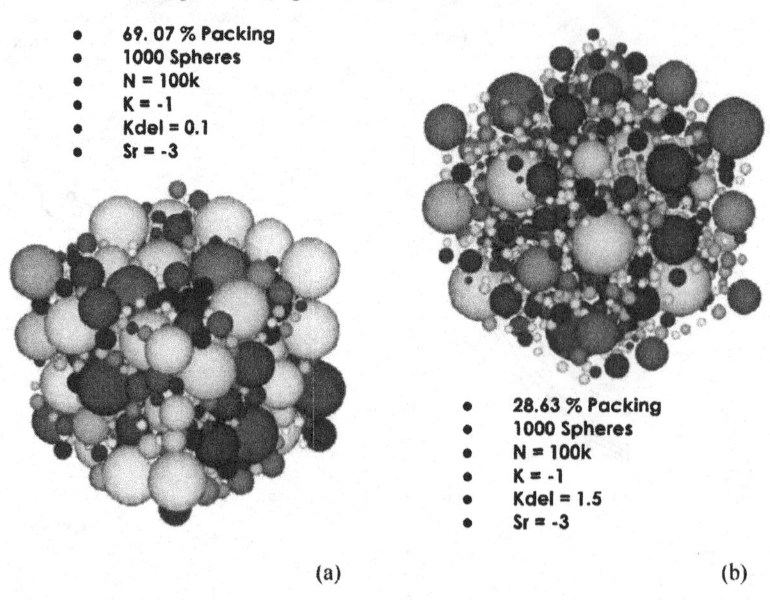

Fig. 7. The application of the model for packing of 1000 spheres at a reduction coefficient -1 and separation coefficient 0.1 (a) and 1.5 (b)

The Effect of the Number of Particles Packed

The effect of an increase in the number of particles packed (experiment B) was investigated for the "best fit" LIP case with reduction coefficient -3 and separation coefficient 1.5. The results of the experiment support the suggestion regarding the fractal nature of particle's packing [11]. Here, the Fuller curves with exponent of 0.45 and 0.7, plotted at a log-log scale, set the boundaries for the aggregate's distributions used in concrete technology. With the increase in the number of particles packed the actual particle size distribution fits the line tangential to the curve at the origin (corresponding to D_{max}), as shown in Fig. 8. With increased number of smaller-sized particles packed, the particle size distribution curve approaches this tangential line. However, the increase in the number of particles up to 10 million does not significantly improve the packing degree.

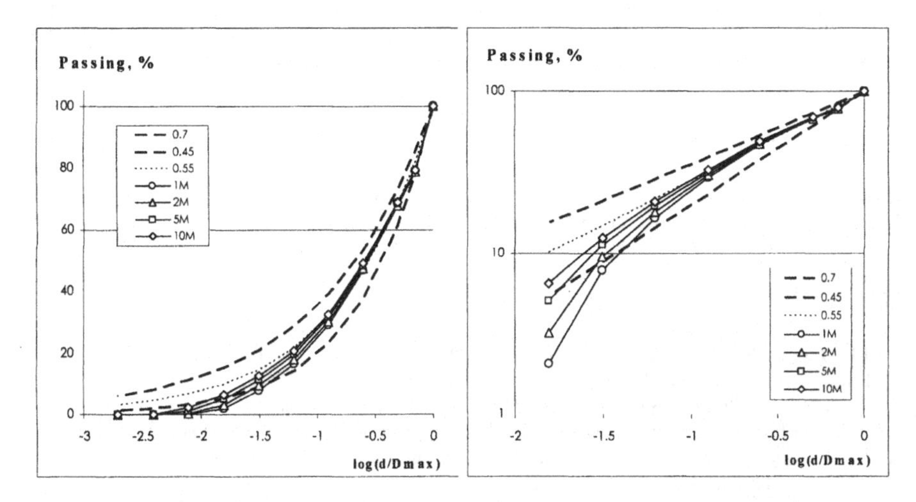

Fig. 8. The effect of the number of particles packed on particle size distribution (reduction coefficient -3)

CONCLUSIONS

1. For a given number of particles, dense packing is achieved either when dense pre-packing is realized or when a wide range of particle sizes is available. The best packing curves with an 85-89% packing degree are presented by the "modified Fuller type", gap or "Initially Pre-Packed" gradings [5]. The characteristic feature of the IPP gradings is related to high (60%) values of packing degree obtained by the narrow range of particle sizes from D_{max} to $D_{max}/2$. The IPP condition occurs when low values of reduction and separation coefficient are set.

2. The opposite case is represented by a Solid Suspension Model characterized by a "Loose Initial Packing" or LIP condition when a high (up to 85 %) packing degree is obtained due to

utilization of a wide range of particle sizes. This condition, represented by the models with a reduction coefficient -3 and separation coefficient 1.5 is characteristic of many "real" particulate assemblies used in concrete.

3. The third case, related to the arrangements of diluted suspensions, is obtained with high values of the reduction coefficient (up to -1) and at a high value of separation coefficient (0.5 and higher). This arrangement can be characterized by high volumes of relatively coarse particles, but with low packing density or concentration of particles.

4. The fractal behavior of particle sizes is observed with an increase in the number of spheres packed (with more fine particles packed). The particle size distribution curve, plotted in log-log scale, fits the line tangential to the curve at the origin (corresponding to D_{max}). With increased numbers of smaller-sized particles packed, the particle size distribution curve approaches this tangential line. However, at this stage, the increase in the number of particles up to 10 million does not significantly improve the packing degree.

ACKNOWLEDGEMENTS

This study was performed under the research grants provided by PROMEP and PAICyT. The financial support of these institutions is gratefully acknowledged.

REFERENCES

1. J.F. Young, S. Mindess, R. J. Gray, A. Bentur, The Science and Technology of Civil Engineering Materials, Upper Saddle River, NJ : Prentice Hall, 384 p.
2. A.M. Neville, Properties of Concrete, Prentice Hall, 2000, 844 pp.
3. D.J. Cumberland and RJ. Crawford, The Packing of Particles, Elsevier, Amsterdam, 1987.
4. V.A. Vorobiev, Application of Physical and Mathematical Methods in Concrete Research, Visshaya Shkola, Moscow, 1977 (in Russian).
5. K. Sobolev and A. Amirjanov, The Development of a Simulation Model of the Dense Packing of Large Particulate Assemblies. Powder Technology, Vol. 141, No. 1-2, 2004, pp. 155-160.
6. H.-G. Kessler, Spheres Model for Gap Grading of Dense Concretes, BFT 11 (1994) 73-75.
7. K. Sobolev and A. Amirjanov, A Simulation Model of the Dense Packing of Particulate Materials. Advanced Powder Technology, Vol. 15, No. 3, 2004, pp. 365-376.
8. W.B. Fuller and S.E. Thompson, The Laws of Proportioning Concrete, ASCE J. Transport. 59 (1907).
9. Sobolev K., The Development of a New Method for the Proportioning of High-Performance Concrete Mixtures. Cement and Concrete Composites, Vol. 26, No. 7, 2004, pp. 901-907.
10. A. Amirjanov and K. Sobolev, Optimal Proportioning of Concrete Aggregates Using a Self-Adaptive Genetic Algorithm. Computers and Concrete, Vol. 2, No. 5, 2005, pp. 411-421.
11. A. Amirjanov and K. Sobolev, Fractal Properties of Dense Packing of Spherical Particles (in review).

12. R Jullien, A Pavlovitch and P Meakin, Random packings of spheres built with sequential models, J. Phys. A: Math. Gen. 25 (1992) 4103-4113.
13. P. Stroeven and M. Stroeven, Assessment of Packing Characteristics by Computer Simulation, Cem. Con. Res. 29 (1999) 1201-1206.
14. P.A. Cundall, Computer Simulation of Sense Sphere Assemblies, Micromechanics of Granular Materials, Ed. by M. Satake and J. T. Jenkins (1988) 113-123.
15. W. Scoppe, Computer simulation of random packings of hard spheres, Powder Technol. 62 (1990) 189-196.
16. G.T. Nolan and P.E. Kavanagh, Computer simulation of random packing of hard spheres, Powder Technol. 72 (1992) 149-155.
17. K.Z.Y. Yen and T.K. Chaki, A Dynamic Simulation of Particle Rearrangement in Powder Packings with Realistic Interactions, J. Appl. Physics 71 (1992) 3164-3173.
18. P.A. Cundall and O.D.L. Strack, A Discrete Numerical Model for Granular Assemblies, Geotechnique 29(1) (1979) 47-65.
19. A.J. Matheson, Computation of a random packing of hard spheres, J. Phys. C: Solid State Phys. 7 (1974) 2569-2576.
20. W.M. Visscher and M. Bolsterli, Random packing of equal and unequal spheres in two and three dimensions, Nature 239 (1972) 504-507.
21. S. Torquato, T. M. Truskett, and P. G. Debenedetti, Is Random Close Packing of Spheres Well Defined? Phys. Rev. Lett. 84 (2000) 2064-2067.
22. S.V: Anishchik and N.N. Medvedev, Three-Dimensional Apollonian Packing as a Model for Dense Granular Systems, Phys. Rev. Lett. 75 (1995) 4314–4317.
23. D. Coelho, J.–F. Thovert, and P. M. Adler, Geometrical and transport properties of random packings of spheres and aspherical particles, Phys. Rev. E 55 (1997) 1959–1978.
24. J.C. Kim, K.H. Auh and D.M. Martin, Multi-level particle packing model of ceramic agglomerates, Modelling Simul. Mater. Sci. Eng. 8 (2000) 159-168.
25. P.J. Andersen and V. Johansen, Particle Packing and Concrete Properties, in Materials Science of Concrete II, J. Skalny and S. Mindess, eds., The American Ceramic Society, Westerville, Ohio. (1995) 111-146.
26. P. Goltermann, V. Johansen and L. Palbol, Packing of Aggregates: an Alternative Tool to Determine the Optimal Aggregate Mix, ACI Mat. J. (1994) 435-443.
27. X. Jia and R.A. Williams, A Packing Algorithm for Particles of Arbitrary Shapes, Powder Technol. 120/3 (2001) 175-186.
28. W.S. Jodrey and E.M. Tory, Computer simulation of isotropic, homogeneous, dense random packing of equal spheres. Powder Technol. 30 (1981) 111-118.
29. J.D. Sherwood, Packing of spheroids in three-dimensional space by random sequential addition, J. Phys. A 30 (1997) L839-L843.
30. K. Sobolev, A. Amirjanov, R. Hermosillo and F.C. Lozano, Packing of Aggregates as an Approach to Optimizing the Proportioning of Concrete Mixtures Aggregates: Asphalt Concrete, Portland Cement Concrete, Bases, and Fines – ICAR/AFTRE/NSSGA Symposium, April 4-7, 2004, Denver, Colorado, USA.

MODELING OF STIFFNESS DEGRADATION AND EXPANSION IN CEMENT BASED MATERIALS SUBJECTED TO EXTERNAL SULFATE ATTACK

B. Mobasher
Department of Civil and Environmental Engineering
Arizona State University, Tempe, AZ, USA

ABSTRACT

Sulfate attack in concrete structures is considered to be among the major durability concerns in civil infrastructure systems. Proper modeling techniques can help us understand the influence of aggressive environments on the concrete performance, and improve the decision making process in every stage of construction and maintenance. Aspects of cement chemistry, concrete physics, and mechanics are applied to develop a diffusion-reaction based model for predicting sulfate penetration, reaction, damage evolution, and expansion, leading to degradation of cement-based materials exposed to a sulfate solution. The model is refined to address the interaction effects of various parameters using calibration data available from an array of experimental results. A simplified approach is presented to compute the rate of degradation and expansion potential using a series solution approach. Model parameters of are obtained through parametric analysis and calibration with experimental data.

INTRODUCTION

Portland cement-based materials subjected to attack from sulfates may suffer from two types of damage: loss of strength of the matrix due to degradation of calcium-silicate-hydrate (C-S-H), and volumetric expansion due to formation of gypsum or ettringite that leads to cracking. Loss of strength has been linked to decalcification of the cement paste hydrates upon sulfate ingress, especially C-S-H, while cracking and expansion is attributed to formation of expansive compounds. Efforts of modeling the durability due to external sulfate attack have received attention in the past decade [1]. An empirical relationship between ettringite formation and expansion is the basis for many models where the expansive strain is linearly related to the concentration of ettringite [2]. This approach has been incorporated in the 4SIGHT program which predicts the durability of concrete structures [3], as well as in a model that calculates the service life of structures subjected to the ingress of sulfates by sorption [4] or mechanical and transport properties [5].

The general governing phenomena for the transfer of mass through concrete is modeled by means of conservation-type equations involving diffusion, convection, chemical reaction, and sorption. In the case of sulfates, some authors [6] assume that the process is controlled by reaction rather than diffusion, based on an empirical linear equation that links the depth of deterioration at a given time to the tri-calcium silicate (C_3A) content and the concentration of magnesium and sulfate in the original solutions. A solution of the diffusion equation with a term for first order chemical reaction was proposed to determine the sulfate concentration as a function of time and space [7,8]. Similar to the recent work by the NIST group, the diffusion coefficient is represented as a function of the capillary porosity and varies with time since capillary pores fill up with the recently formed minerals [4]. Using micromechanics theory and the diffusion-reaction equation, a model that predicts the expansion of mortar bars has been developed for the 1-D case [9].

A chemo-mechanical mathematical model developed recently simulates the response of concrete exposed to external sulfate solutions [10,11]. The model is based on the diffusion-reaction moving boundary approach and several mechanisms for the reaction of calcium aluminates with sulfates to form expansive ettringite are considered. There are three major input parameters categorized under the main categories of 1) Material Parameters, 2) Exposure & Environmental Loading, and 3) Size & Shape of members. Input parameters are used to estimate physical parameters such as the diffusivity, strength, concentration of available calcium aluminates, and the volumetric proportions due to chemical reactions.

Three distinct but coupled problems of sulfate diffusion, calcium aluminate depletion, and crack front propagation are treated as a moving boundary problem as shown in Figure 1. As the time parameter increases, the sulfates diffuse, and then react with aluminates, resulting in hydration products which expand and potentially cause cracking. The cracking causes the coefficient of diffusivity to change from an uncracked material D_2 to a cracked material D_1. This change may be linked to a scalar damage parameter that also affects the material stiffness, E. This damage parameter, ω, is defined from the available models for uniaxial stress-strain response [10].

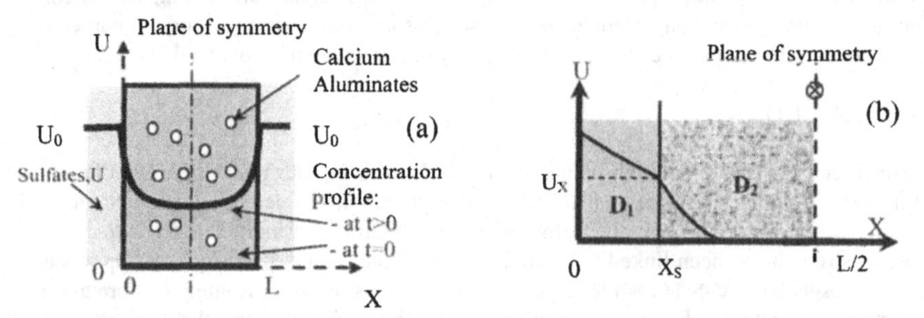

Figure 1. a) Sulfate concentration profile in a specimen of length L at two instances t=0 and t>0. b) The variation of concrete diffusivity as a function of crack front located at $X=X_s$

It is assumed that the calcium aluminates may be a blend of three different phases of tricalcium aluminate, tetracalcium alumino hydrate, and monosulfate with parameter γ_i representing the proportion of each phase. The cement chemistry notation is used with (C=CaO, S=SiO$_2$, A=Al$_2$O$_3$, H=H$_2$O, and $\bar{S}$=SO$_3$). The total calcium aluminate phase is then introduced as: "C_a" ($C_a = \gamma_1\ C_4AH_{13} + \gamma_2\ C_4A\bar{S}H_{12} + \gamma_3\ C\bar{S}H_2$). Each of these compounds may react with the ingressing sulfates (represented in the form of gypsum) according to stoichiometric amounts defined in equations 1-3:

$$C_4AH_{13} + 3C\bar{S}H_2 + 14H \longrightarrow C_6A\bar{S}_3H_{32} + CH \tag{1}$$

$$C_4A\bar{S}H_{12} + 2C\bar{S}H_2 + 16H \longrightarrow C_6A\bar{S}_3H_{32} \tag{2}$$

$$C_3A + 3C\overline{S}H_2 + 26H \longrightarrow C_6A\overline{S}_3H_{32} \qquad (3)$$

These reactions are lumped in a global sulfate phase-aluminate phase reaction as:

$$C_a + q\overline{S} \longrightarrow C_6A\overline{S}_3H_{32} \qquad (4)$$

where $q\overline{S} = (3\gamma_1 + 2\gamma_2 + 3\gamma_3)C\overline{S}H_2$ represents the weighted stoichiometric coefficients of the sulfate phase. For any of the individual reactions described above, the volumetric change due to the difference in specific gravity can be calculated. The concentrations of the aluminate and the sulfate phases are represented as two parameters U and C.

$$U = U_{SO_4} \text{ and } C = U_{CA} \qquad (5)$$

The coupled differential equation for the penetration of sulfates and their reaction with the calcium aluminate phase is represented as a first order diffusion reaction equation and represented as:

$$\frac{\partial U}{\partial T} = D\frac{\partial^2 U}{\partial X^2} - kUC \qquad (6)$$

$$\frac{\partial C}{\partial T} = -\frac{kUC}{q} \qquad (7)$$

Equation 6 represents the rate of change of concentration of sulfates as a function of sulfate diffusivity and also the rate of reaction of sulfates presented as the parameter k. Equation 7 represents the rate of reaction of Aluminate phase and since they are assumed to be stationary, no diffusion of this phase considered. Therefore, the change of aluminates is considered as a function of time only. The general solution for equations 6 and 7 are presented using a finite difference method by Tixier and Mobasher [10,11]. The coupled differential equations for the depletion of both sulfates and aluminates are solved by means of numerical techniques to take into account the three main effects: a) limited supply of C_3A, b) cracking induced changes in diffusivity, and c) the degradation effects of the expansive equations.

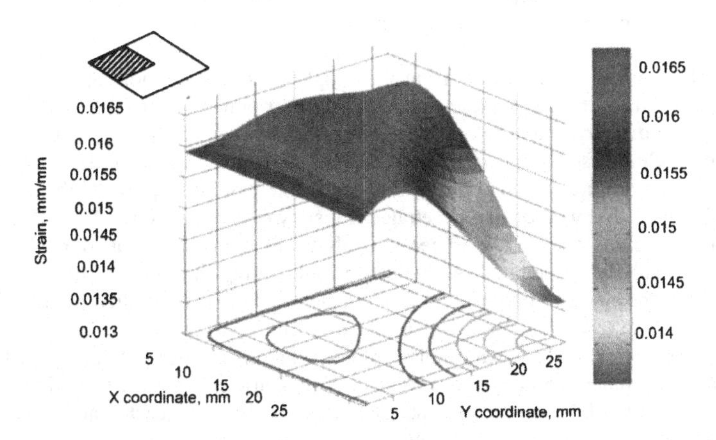

Figure 2. Strain generation across a 1/4 portion of the sample exposed on both sides

Figure 2 represents the simulation of the diffusion reaction equation and the strain generation across a 1/4 portion of the sample exposed on both sides. The model predicts the amount of reacted aluminates, the stresses generated, and internal parameters such as damage, expansion levels, weight gain, stiffness degradation and tensile strength degradation [11].

It is possible to obtain a series based solution by making simplifying assumptions regarding the interaction of Sulfates and the Aluminate phases. If we assume there is sufficient amount of aluminates present so that there is no depletion of this phase, one can correlate the rate of reaction of the Sulfates with the Aluminate phase into a single material constant, *k*, representing the rate of depletion of sulfates. The higher this value the more readily available the reaction of aluminates will be. Furthermore it is assumed that the diffusivity remains constant and is not affected by the cracking (i.e. $D_1 = D_2$). Equation 6 can be simplified as a single variable second order partial differential equation represented as Equation 8 and its series solution is represented as equation 9.

$$\frac{\partial U}{\partial T} = D\frac{\partial^2 U}{\partial X^2} - kU \tag{8}$$

$$\frac{U}{U_0} = 1 - \frac{4}{\pi}\sum_{m=0}^{\infty}\frac{1}{n(k+\nu)}\sin\left(\frac{n\pi X}{L}\right)\left(k+\nu\exp[-T(k+\nu)]\right) \tag{9}$$

$$n = 2m+1 \quad , \quad \nu = D\left(\frac{n\pi}{L}\right)^2 \tag{10}$$

This simplified approach does not require a finite difference solution for the sulfate penetration and its solution according to equation 9 is referred to as the series solution approach. The sulfate penetration profile throughout the thickness of the sample as a function of diffusivity of the matrix phase after 30 weeks of exposure is shown in Figure 3. The effect of the diffusivity on the concentration profiles as predicted by Fick's second law with a second order reaction are shown. Effect of rate of reaction on the sulfate concentration profiles after a 30 week exposure are shown in Figure 4. Note that when there is no reaction of sulfates taking place, more are able to penetrate inside the material. However, as the aluminates are depleted due to the reaction, then the value of *k* decreases and approaches toward zero, resulting in accumulation of sulfates in the material.

Computation of expansion in a sample using the simplified series solution algorithm is as follows. The solution for the series expansion of sulfate distribution is obtained as a function of time for a given diffusivity and rate of reaction as shown in figure 4. (D, and k= finite amount). The baseline case for the same expression is also solved using an assumption of k=0 (representing no Sulfate depletion due to reaction). The baseline represents the case for the amount of sulfates that would have penetrated if there were no reactions. The difference between the two levels is the amount of reacted sulfates as defined by Equation 11. The average values of sulfates penetrated is obtained by integrating the U/U_0 values over the entire depth of the specimen. This level (average value of U/U_0) is multiplied by the initial sulfate concentration at the surface (ex. $U_0 = 0.35$ moles/Lit, input) to find the concentration of reacted sulfates. Stoichiometric and molar volume relations are used to convert reacted sulfates to reacted aluminates, and ettringite formed. Volumetric changes are related to linear expansion values. The total calcium aluminate phase is divided into reacted and unreacted amounts and represented respectively as C_{ar} and $C_{au}(x,t)$ according to:

$$C_{ar}(x,t) = C_a - C_{au}(x,t) \tag{11}$$

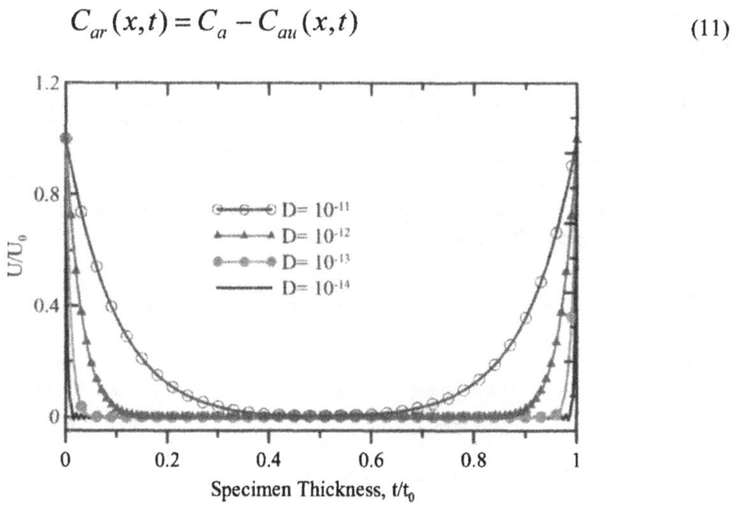

Figure 3. Sulfate penetration profile as a function of matrix diffusivity 30 weeks exposure.

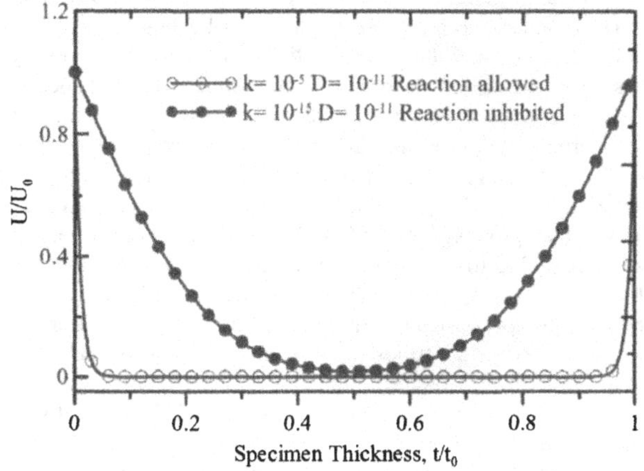

Figure 4. The sulfate penetration profile as a function of position after 30 weeks, effect of reaction rate on the sulfate concentrations are shown

Using a rule-of-mixtures approach, one can relate the expansive nature of the products with the prescribed specific gravity of the compounds. In the present approach, calculations of the volumetric changes between reactants and products were conducted by assuming that ettringite was the only product obtained. Formation of other products such as AFM phase or Gypsum could be also addressed in the volumetric calculations. Once the amount of reacted calcium aluminates into ettringite as a function of time and space are obtained, they can be related to the volumetric strain and the volume changes. It is furthermore assumed that the crystallization pressure of products of reaction results in a bulk expansion of the solid. The constitutive response of the matrix and the expansive stresses are calculated from the imposed volumetric strain. In the main model microcracks are initiated when the strength of the matrix is reached, leading to changes in the diffusivity and a reduction in matrix elastic properties. The variation of diffusivity is linked to the scalar damage parameter due to cracking of the matrix. In the simplified model, the expansive strains are used in the constitutive stress crack width response and an elastic equivalent stiffness is defined to account for the reduced stiffness of the sample. No changes in diffusivity are considered in the simplified model. An averaging scheme is used the corresponding expansion based on the molar volumes is defined as:

$$\varepsilon_V^t(x,t) = \varepsilon_V^0(x,t) - f\,\Phi = C_{ar}\sum_P\left(\frac{\Delta V}{V}\right)_P - f\,\Phi \qquad (12)$$

According to equation 12, the volumetric strain is directly proportional to the volumetric change due to the reaction products and adjusted by a shift factor representing the total capillary porosity Φ. Parameter f is defined as the fraction of capillary porosity available for the dissipation of the expansion products. The magnitude of the shift (delay) in the expansion is due to the amount of capillary porosity.

As the expansive pressures generate, localized tensile stresses are imposed on the internal pore structure. These tensile stresses are assumed to be applied to a sample under end restrain and cause cracking and thus degradation. In order to properly formulate this response, one must represent the crack formation and propagation and specifically the closing pressure profile that normally exists in cement based materials. In the present case, it is assumed that the internal expansive pressures relieve the generation of closing pressures in the vicinity of cracks. The toughening behavior which is an inherent component of failure in cement based materials is therefore lost due to the existence of internal pressures due to the chemical reactions. The proposed procedure is based on a simplified bridging tractions formulation. The first step is to utilize a stress crack width relationship. A simple uniaxial tensile stress-strain law proposed by Sakai and Suzuki [12] is used to represent the constitutive response and represented in Equations 13 and 14. The material is assumed to be linear elastic up to the peak stress, following a strain softening rule for the descending part. Similar in nature to Foote, Mai, Cotterell, [13], this approach represents the stress across the crack ligament as a function of both the crack opening and the crack ligament length. By assuming various functional relationships, models of the decreasing stress as a function of crack opening are represented. For example, the responses for both stress crack opening and crack opening vs. position can be expressed as equations 13 and 14 respectively. Parameter l_b in this case is equivalent to the stable crack growth length Δa_c.

$$\sigma_b = \sigma_b^0\left[\left(\frac{x}{l_b}\right)^q\right]^{n_d} \qquad (13)$$

$$u_b(x) = u_b^0\left(\frac{x}{l_b}\right)^n \qquad (14)$$

The deformation at peak stress is obtained as: $w_0 = \varepsilon_p$ x H, where H is the gage length of the specimen, and ε_p is the strain at peak. The post peak response is assumed to be by means of a power curve with its coefficients defined as n, q, and n_d. In the present model, effect of shape of tensile stress crack opening profile was not studied and constant parameters representing $w_0 = 0.38$, $q = 0.5$, and $n_d = 1.5$ were used. The only variable of this model was the tensile strength parameter, $f'_t = \sigma^0_b$. It was furthermore assumed that the unloading is through an elastic recovery to strain at ultimate strength and the only inelastic displacement is the deformation at peak stress. The modulus in the post-peak region is obtained as the average of the points within the damaged zone and undamaged zone:

$$E = \frac{\sigma_b}{\varepsilon - \varepsilon_p} \qquad \bar{E} = \frac{1}{A}\int_A E\,dA \tag{15}$$

For a prismatic specimen subjected to sulfate ingress from all sides, a uniaxial condition is used, and the stiffness is averaged across the cross section, A, using equation 15. Furthermore, it is assumed that normal strain is the primary mode of deformation and no curvature is induced throughout the cross section. The averaged expansion is obtained using an averaging algorithm:

$$\Delta = \sigma_r\left(\frac{1}{\bar{E}} - \frac{1}{E_0}\right)H \tag{16}$$

Parameter $\bar{E}$ is the average instantaneous modulus over the cross-section. Parameter σ_r is assumed to be a constant uniform residual stress in the specimen due to past loading history prior to sulfate attack (i.e. a uniform shrinkage stress). The residual stress level in the present approach is assumed as 2-10 MPa, and may be viewed as the scaling parameter used to relate the changes in the stiffness of the sample to the expansion levels observed in experiments.

MODEL VERIFICATION

Several experimental observations are explained in the framework of the present mathematical model. Two different types of studies are conducted which include parametric evaluation and calibration of the experimental data. Parametric study of the effect of initial sulfates and the effect of cracked material diffusivity on the expansion characteristics of the sample are presented first. The effect of initial sulfate phase concentration on the diffusion and subsequent degradation mechanisms, in addition to the effect of degree of hydration, is studied. The experimental verification evaluates the response of specimens with blended cements and various Calcium Aluminate levels in addition to a simulation of degradation of elastic properties.

EFFECT OF CHANGES IN DIFFUSIVITY DUE TO USE OF MINERAL ADMIXTURES

It is generally accepted that the use of mineral admixtures such as silica fume, flyash or slag significantly improves the sulfate resistance of concrete. This is verified through experimental tests of various researchers [14, 15]. A set of parametric studies were conducted to show that by changing the diffusivity of the base material and the extent of cracking induced diffusivity changes, one can address the potential benefits of the supplementary cement based materials.

A variable diffusivity parameter was introduced in order to reflect the use of mineral admixtures in changing the microstructure. Test data comparing samples with two different initial diffusivities of $D_2 = 1\times10^{-12}$ m^2/s and 1×10^{-13} m^2/s were considered in Figures 5a and 5b. In each of these trial runs, a range of diffusivity changes due to cracking was also considered spanning three orders of magnitude. A parameter $D_1/D_2 = 10$ would indicate that the cracked material would increase its diffusivity by ten fold as compared to the uncracked base material. Results are shown in Figures 5a and 5b and indicate that as the time increases, an increased diffusivity due to cracking would directly result in faster rates of expansion. The expansion and degradation would take place faster by increasing the D_1/D_2 ratio. This response also happens much faster for the specimen with a higher initial diffusivity (D_2= $1x10^{-12}$ m^2/s). Analysis of various experimental results indicates that the choice of the range of values for D_2 (undamaged material diffusivity) is consistent with the water/cement mass ratio and microstructure of the mix designs used. With all other parameters were kept the same, lower values of w/c and the use of pozzolanic materials led to a lower diffusivity, and hence to a slower reaction and expansion rate. This mechanism is clearly predicted by the model. When the uncracked diffusivity is reduced by an order of magnitude from $1x10^{-12}$ m^2/s to $1x10^{-12}$ m^2/s the period to achieve the same level of expansion is significantly extended as well. This time factor depends on the level of damage caused by the cracking as well. Note that the higher the level of damage due to cracking as measured by D_1/D_2 ratio, the faster the degradation process. The linear strain measure was used to represent the expansion term here. It was assumed that the linear expansion component would be 1/3 of the volumetric strain under the conditions that the strain is homogeneous and isotropic.

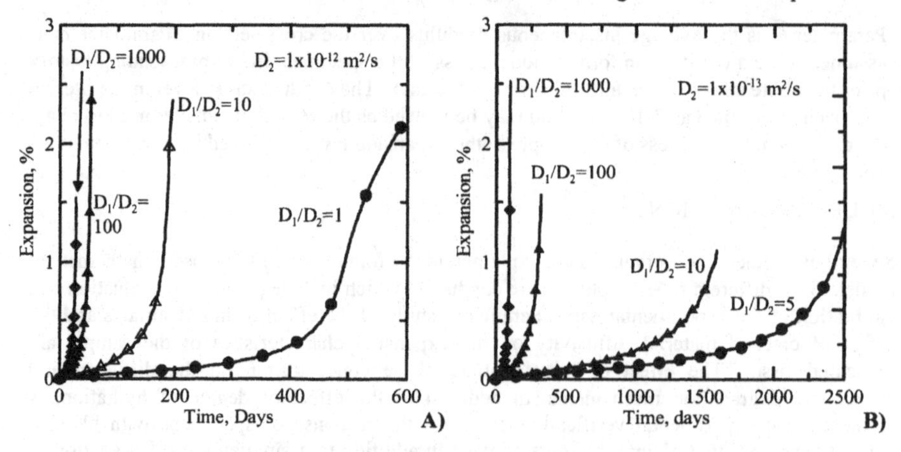

Figure 5 Parametric study of the effect of diffusivity of the cracked material on the expansion-time curves for specimens with various levels of damage due to cracking.

Figure 6 presents the simulation of the expansion results of a mixture containing 20% class F flyash with a control sample. The w/c ratio of both mixtures is 0.485, and the C_3A content of the Portland cement used was 6.3%. Note that as the cement is replaced with flyash, the total C_3A content is reduced proportionately. A tensile strength of f_t= 2 MPa, and a residual stress of σ_r = 10 MPa were used in these simulations. Note that both the experiments and simulations show a significantly reduced risk of expansion potential with the use of flyash. In order to fit the experimental data in the present model, a relatively homogeneous initial diffusivity ranging from D_2= $5x10^{-13}$ for control to D_2=$7x10^{-13}$ for flyash mixtures were used. Note that initially the two samples are behaving similar to one another. However, the most important model parameter as compared to the parametric study is the amount of change in the diffusivity due to the cracking. In the case of samples with flyash a ratio of D_1/D_2=3 as compared to a value of 10 for the control sample is used. This model indicates that the damage due to the Sulfate attack of a specimen containing flyash is less detrimental than the damage in a control specimen.

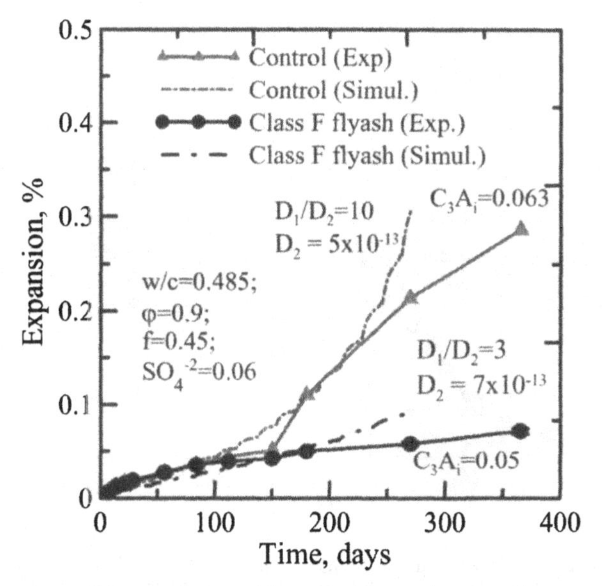

Figure 6. Comparison of the simulation with expansion results of a mixture containing Class F flyash with a control sample

Figure 7 presents the simulation of the expansion results of a mixture containing 40% high CaO flyash with a tertiary mixture containing 3% silica fume mixture. Test results were obtained from the experiments of Thomas et. al [16]. The w/c ratio of both mixtures is 0.45 and the C_3A=6.8% was used for the Portland cement. A tensile strength of f'_t= 3 MPa, and a residual stress of σ_r = 10 MPa were used in these simulations. Once again it is observed that both the experiments and simulations show a significantly reduced risk of expansion potential with the use of tertiary blends. In order to fit the experimental data in the present model, the series solution was used and the diffusivity from $D_2=8.0x10^{-13}$ and $D_1/D_2=5$ for control to $D_2=2.0x10^{-13}$ and $D_1/D_2=5$ for flyash mixtures were used. Note that initially the two samples are behaving similar to one another. The experimental data can also be explained using the series solution which utilizes a single diffusivity parameter. In this manner, the most important model parameter is the amount of change in the diffusivity of the two mixtures. This model indicates that even with high CaO flyash mixtures, use of a tertiary blend can help with the diffusivity of the mixtures. The diffusivity values used for the series solution fits were $D=1.14x10^{-11}$ and $D=1.5x10^{-12}$ for control and blended cements respectively.

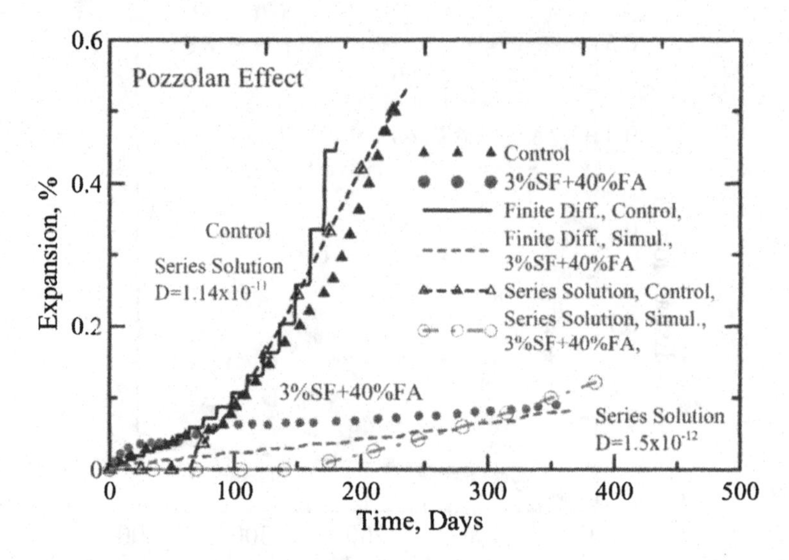

Figure 7. Comparison of the simulation with expansion results of a mixture containing Class F flyash and Silica fume with a control sample

EFFECT OF INITIAL GYPSUM CONTENT

Effect of initial gypsum content on the expansion-time curves for specimens containing two levels of initial C_3A, 6 % and 10 %, are presented in Figures 8a, and 8b. Note that in both cases, increase in the amount of initial gypsum increases the potential for damage and degradation. It is interesting to note that the rate of expansion is different in different samples as well. In specimens containing 10 % initial C_3A, even with only the use of 2 % initial sulfates, significant damage is observed after 120 d (0.6%), whereas with the same initial sulfates content, a specimen with 6 % initial C_3A would not show a significant expansion within that same period.

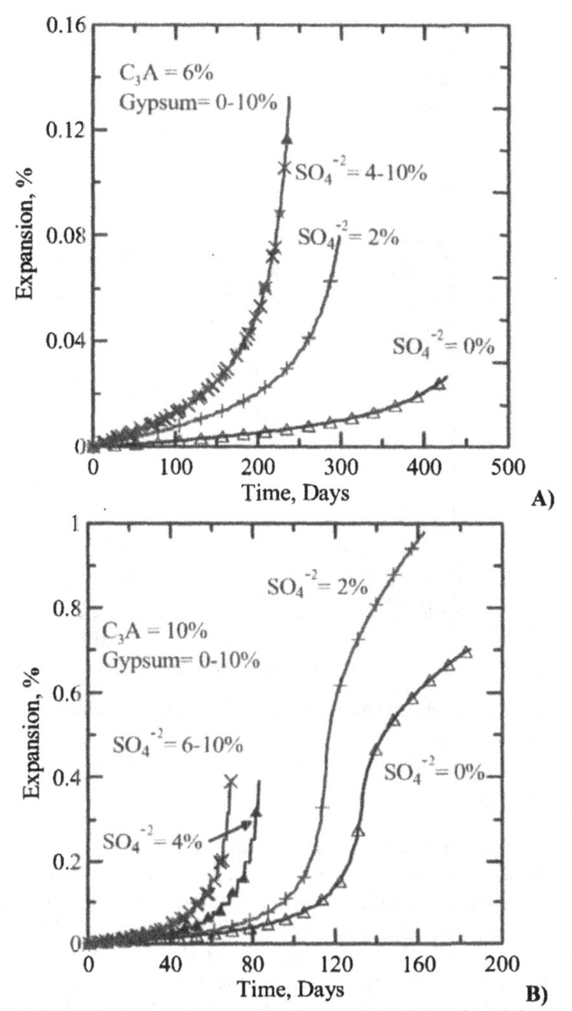

Figure 8 Effect of initial gypsum content on the parametric study of the expansion time curves for specimens containing two levels of initial C_3A, A) C_3A= 6 %, and B) C_3A=10 %.

EFFECT OF INITIAL C_3A CONTENT

The resistance of plain and blended concrete (C_3A content between 4.3% - 12 %) to sulfate solutions has been studied by several authors [17]. Effect of initial C_3A content was studied by evaluating the test results of a series of experiments conducted by Ferraris et al. [18]. A reasonable fit is obtained by adjusting the input data in this case. The results are shown in Figures 9. The values of the parameters for fitting the data are reported as a degree of hydration of 0.9, and an initial sulfate content equivalent to 6% was assumed. Note that as the initial C_3A increases, the expansion also increases as a function of time. Figure 9a shows a set of data obtained for specimens with an initial C_3A content of 7 and 12 %. Note that the rate of expansion in samples with higher tricalcium aluminates is significantly higher than the control case, and the simulation graphs are able to address this enhanced expansion activity quite well. Due to the large magnitude of expansive forces expected in this case, the amount of residual stress assumed in this case was set at 10 MPa, as compared to the previous case, which was assumed at 5 MPa.

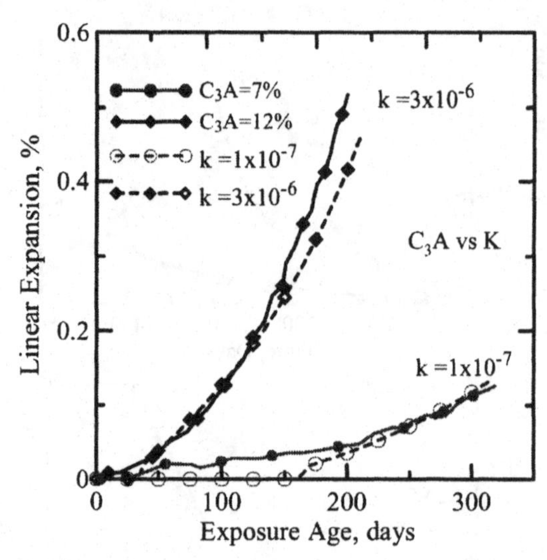

Figure 9 Effect of initial C_3A content on the expansion time response of specimens. [19]

Simulation of degradation

In order to simulate the average degradation in a sample, the test results of the average value of the Young's modulus across the cross section of the specimen is plotted in Figure10. Test results compare the expansion vs. time profiles of experiment vs. the simulated curves. The average reduction of the young's modulus was obtained using dynamic modulus test equipment as a function of exposure time. The simulated values of average young's modulus were obtained using equation 15 which averaged the effective young's modulus throughout the cross section. Note that during the first 30 days after subjecting the sample to the sulfate solution, the stiffness of the sample increases and then after reaching a plateau it decreases. This may be attributed to the increased volume of hydration products filling up of the pores, and therefore reducing the internal porosity. Any excess filling up of the pores, results in generation of internal pressures which ultimately lead to degradation and reduction of stiffness of the sample. Note that the present model predicts a rate of degradation which is quite consistent with the experimental observations.

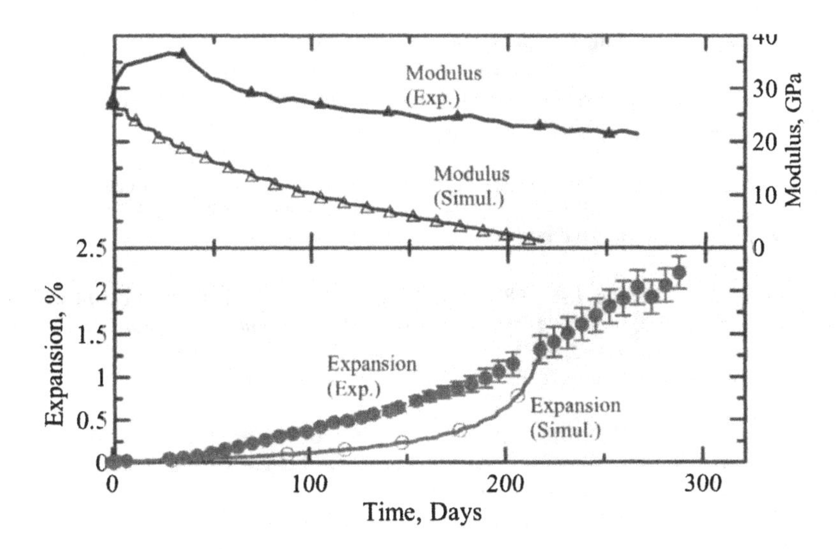

Figure 9. Correlation of Expansion results with modulus degradation.

CONCLUSIONS

Applications of a theoretical simulation model to predict the degradation due to external sulfate attack on cement-based materials are discussed. Simulations of the model using a series of parametric studies indicate that the effects of diffusivity of the cracked and uncracked material can play a significant role in the characteristics of sulfate penetration. The amount of initial sulfates in the Portland cement is also an important parameter as predicted by the model. A model based on series solution was presented and while the model uses a single diffusivity parameter, it is capable of predicting the test experimental test results using average properties. Parameters of the model are obtained using the actual mix design of the materials. Effect of initial C_3A content and the sulfate concentration during testing were shown to be predicted favorably by the model parameters.

ACKNOWLEDGEMENTS

The financial support of the Salt River Project, Phoenix, Arizona for the research program on “Blended Cements” is greatly appreciated..

REFERENCES

1 Philip, J. and Clifton, J.R (1992). “Concrete as an Engineered Alternative to Shallow Land Disposal of Low Level Nuclear Waste: Overview.” *Fly Ash, Silica Fume, Slag, and Natural Pozzolans in Concrete. Proceedings – Fourth International Conference.* Ed. V.M. Malhotra. Detroit: American Concrete Institute, vol. 1, p.713-730.

2 Atkinson, A; Haxby, A. and Hearne, J.A. (1988) “The Chemistry and Expansion of Limestone-Portland Cement Mortars Exposed to Sulphate-Containing Solutions.” *NIREX Report NSS/R127*, United Kingdom: NIREX.

3 Snyder, K.A., Clifton, J.R. and Pommersheim, J. (1995). "Computer Program to Facilitate Performance Assessment of Underground Low-Level Waste Concrete Vaults." *Scientific Basis for Nuclear Waste Management XIX*. Ed. W.A. Murphy and D.A. Knecht. Pittsburgh, Pa.: Materials Research Society, p. 491-498.

4 Bentz, D.P.; Clifton, J.R.; Ferraris, C.F. and Garboczi, E.J. (1999) "Transport Properties and Durability of Concrete: Literature Review and Research Plan." *NISTIR 6395*. Gaithersburg, MD, NIST.

5 Atkinson, A., Hearne, J.A (1989). "Mechanistic Model for the Durability of Concrete Barriers Exposed to Sulfate-Bearing Groundwaters." *Scientific basis for nuclear waste management XIII.* Ed. V.M. Oversby and P.W. Brown. Pittsburgh, Pa.: Materials Research Society, p. 149-156.

6 Pommersheim, J.M., Clifton J.R (1994)."Expansion of Cementitious Materials Exposed to Sulfate Solutions." *Scientific Basis for Nuclear Waste Management. Materials Research Society Symposium Proceedings XVII*. Ed.A. Barkatt and R. Van Konynenburg. Pittsburgh, Pa. : Materials Research Society, p. 363-368.

7 Gospodinov, P., Kazandjiev, R., and Mironova, M. (1996). "Effect of Sulfate Ion Diffusion on the Structure of Cement Stone." *Cement & Concrete Composites* 18 (6), p. 401-407.

8 Gospodinov, P.N, Kazandjiev, R.F., Partalin, T.A. and Mironova, M.K.(1999). "Diffusion of Sulfate Ions into Cement Stone Regarding Simultaneous Chemical Reactions and Resulting Effects." *Cement and Concrete Research* 29 (10), p. 1591-1596.

9 Krajcinovic, D, Basista, M., Mallick, K.and Sumarac, D(1992)."Chemo-Micromechanics Of Brittle Solids." *Journal of the Mechanics and Physics of Solids* 40 (5), p. 965-990.

10 Tixier, R., Mobasher, B., (2003) "Modeling of Damage in Cement–Based Materials Subjected To External Sulfate Attack- Part 1: Formulation", *ASCE Journal of Materials Engineering*, 15 (4), p. 305-313.

11 Tixier, R., Mobasher, B., (2003) "Modeling of Damage in Cement–Based Materials Subjected To External Sulfate Attack- Part 2: Comparison with Experiments, *ASCE Journal of Materials Engineering*, 15 (4), p. 314-322.

12 Suzuki, T., Sakai, M., "A Model For Crack-Face Bridging", International Journal Of Fracture 65 (4): 329-344 Feb, 15, 1994.

13 Foote, R.M.L., Mai Y.W, Cotterell, B., "Crack-Growth Resistance Curves In Strain-Softening Materials", J.of the Mechanics and Physics of Solids, 34 (6): 593-607 1986.

14 Hester, J. A. (1967) "Fly Ash in Roadway Construction," Proceedings of the First Ash Utilization Symposium. U.S. Bureau of Mines, Information Circular No. 8348, Washington, DC, p. 87-100.

15 Dunstan, E. R., Jr. (1980) "A Possible Method for Identifying Fly Ashes that will Improve Sulfate Resistance of Concrete," *Cement, Concrete and Aggregates*, 2(1), ASTM.

16 Thomas M.D.A., Shehata M.H., Shashiprakash S.G. , Hopkins D.S., Cail K., "Use of ternary cementitious systems containing silica fume and fly ash in concrete ," Cement and Concrete Research 29 (1999) 1207–1214

17 R.E. Rodriguez-Camacho, (1998) "Using Natural Pozzolans to Iimprove the Sulfate Resistance of Cement Mortars", Fly Ash, Silica Fume, Slag, and Natural Pozzolans in Concrete. Sixth CANMET/ACI/JCI Conference in Bangkok, p. 1021-1040.

18 Ferraris C.F., Clifton J. R., Stutzman P.E., Garboczi E.J., (1997) "Mechanisms of Degradation of Portland Cement-Based Systems by Sulfate Attack", Proc. of MRS Nov. 1995, Mechanisms of Chemical Degradation of Cement-Based Systems, Ed. K.L. Scrivener and J.F Young, p. 185-192.

19 Ouyang, C. S., Nanni, A., and Chang W. F. "Internal and external sources of sulfate ions in portland cement mortar: two types of chemical attack" Cement and Concrete Research, Volume 18, Issue 5, September 1988, Pages 699-709.

Testing

MEASURING PERMEABILITY AND BULK MODULI USING THE DYNAMIC PRESSURIZATION TECHNIQUE

Z.C. Grasley[1], D.A. Lange[1], G.W. Scherer[2], J.J. Valenza[2]

[1] Newmark Civil Eng. Bldg., 205 N. Mathews Ave., Univ. of Illinois at Urbana-Champaign, Urbana, IL 61801
dlange@uiuc.edu, grasley@uiuc.edu
[2] Civil & Env,. Princeton Univ., Eng./PRISM, Eng. Quad. E-319, Princeton, NJ 08544
scherer@princeton.edu, jvalenza@princeton.edu

ABSTRACT

Novel experimental techniques for measuring the permeability, k, the bulk modulus of the solid phase, K_S, and the bulk modulus of the porous body, K_p have been developed for cement paste and concrete. In the first technique, a saturated sample of concrete or paste is immersed in water and subjected to a sudden increase in pressure, p_A; the sample contracts, then gradually re-expands as the pore pressure equilibrates with the surrounding fluid pressure. The final measured strain depends on K_S. The kinetics of the retardation of the strain is related to the material permeability. The second technique involves applying a hydrostatic pressure to a sealed, partially saturated specimen, such that the measured strain depends on K_p. The first technique has been applied to cement paste specimens of *w/c* 0.4, 0.5, and 0.6. The calculated permeability of the porous body and the bulk modulus of the solid phase for the materials tested compare well with previously reported values for the same materials. The second technique has been applied to cement pastes with *w/cm* ranging from 0.32 to 0.35.

Introduction

The merits of a convenient measurement system for the permeability of cement paste and concrete are readily apparent. Conventional techniques for measuring the permeability of cement paste or concrete have many noted shortcomings. Long test durations (often weeks) require some consideration of material aging. The high pressures required to develop and maintain a constant flow rate through the specimen often initiate leaks through seals, resulting in exaggerated permeability coefficients. A large variability in test data may be a consequence of these limitations [1].

The elastic bulk modulus of the solid microstructure, K_s, and the elastic bulk modulus of the porous cementitious body, K_p, may be quite useful in modeling bulk material response due to microscale stresses. For example, early drying shrinkage and self-desiccation shrinkage are believed to occur primarily as a result of changes in pore fluid pressure associated with the formation of capillary menisci [2]. Underpressure (analogous to tension) in the pore fluid is balanced by corresponding compression in the solid microstructure [3]. Models incorporating the elastic K_s and K_p have been developed for modeling deformations associated with changes in pore fluid pressure [4-5].

The experimental methods described in this paper, which involve a dynamic pressurization (DP) under hydrostatic conditions, allow a relatively rapid and simple determination of k, K_s, and K_p. For a more in depth summary of the measurement techniques for K_s and k, see [6-7].

EXPERIMENTAL

Both hardened cement paste and concrete were tested using the DP technique. The test apparatus involved a pressure chamber that contained the cylindrical specimen and an electric hydraulic pump to apply the hydrostatic pressure. The applied pressure was controlled with an inline regulator, and the pressure was measured using an inline sensor. The axial deformation of the cylinder was measured with an embedment strain gage. Figure 1 illustrates the experimental apparatus.

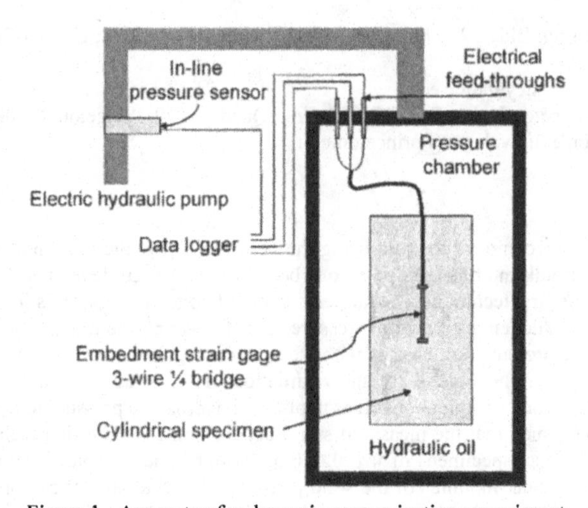

Figure 1. Apparatus for dynamic pressurization experiment.

Two different tests were performed using the apparatus:

- Test 1 measured k and K_s.
- Test 2 measured K_p.

Both tests involve rapidly applying a hydrostatic pressure, then measuring the axial deformation response of the cylindrical specimen. For Test 1 the specimen is saturated prior to testing and allowed to contact the saturated limewater applying the hydrostatic pressure. For Test 2 the partially saturated specimen is sealed to prevent fluid from flowing into the specimen under hydrostatic pressure.

When Test 1 is performed, the specimen initially contracts, then gradually re-expands with time as the internal pore fluid pressure slowly equilibrates with the externally applied pressure. The final strain represents the compression of the solid skeleton of the material under the applied hydrostatic pressure, and is therefore related to K_s. The rate of re-expansion after pressure application is related to the rate of fluid flow in the specimen, and therefore the permeability, k.

When Test 2 is performed, the specimen is sealed to prevent fluid intrusion such that the initial response of the specimen to the applied hydrostatic pressure is related to K_p. In our tests, we have been most successful at preventing leaks when sealing the specimen with a thin layer of epoxy. If the pressure is maintained constant with time, the specimen will slowly creep (increase in compressive strain with time). Figure 2 compares the specimen preparation and expected deformation response of the two different tests.

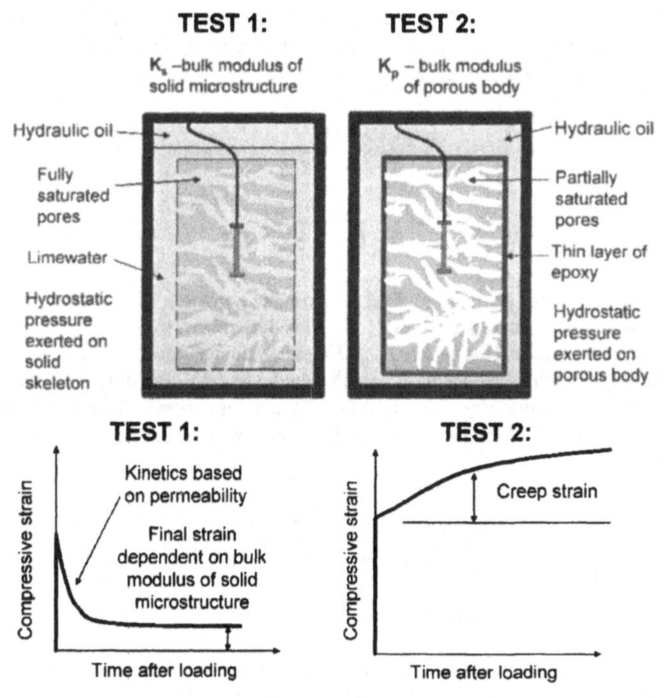

Figure 2. Comparison of Test 1 and Test 2 experimental setup (a.) and typical measured deformation response (b.).

ANALYZING TEST 1

The fluid flow created in Test 1 is assumed to be governed by Darcy's Law, which states that the flux, J, is proportional to the gradient of the pressure, p, such that

$$J=-\frac{k}{\eta_L}\nabla p\,, \tag{1}$$

where k is the intrinsic permeability of the material, and η_L is the viscosity of the pore fluid. The actual retardation response associated with this flow is complicated and requires a numerical inversion of a Laplace Transform [7]. However, the retardation response of a cylindrical specimen exposed to a step function hydrostatic pressure may be described quite accurately by [7]

$$\varepsilon_z = (\varepsilon_\infty - \varepsilon_0)\exp\left\{\frac{4}{\sqrt{\pi}}\left[1 - b\lambda(1-\beta)\right]\left(\frac{(t/\tau_v)^{2.2} - (t/\tau_v)^{1/2}}{1-(t/\tau_v)^{0.55}}\right)\right\} \tag{2}$$

where ε_z is the axial strain, ε_∞ is the final strain, ε_0 is the instantaneous strain after pressurization, $b = 1 - K_p/K_s$ is the Biot coefficient, $\beta=(1+\upsilon_p)/[3(1-\upsilon_p)]$ (where υ_p is the Poisson ratio of the porous body), and the term λ is expressed as

$$\lambda = \frac{Mb}{(Mb^2 + K_p)} \qquad \frac{1}{M} = \frac{\phi}{K_L} + \frac{b-\phi}{K_S} \tag{3}$$

where K_L is the bulk modulus of the pore fluid, and ϕ is the porosity. The porosity is easily measured, K_p may be determined from Test 2, and K_s is determined from the final strain and the applied hydrostatic pressure in Test 1. Errors in K_p and K_s do not significantly affect the calculation since the terms $1/K_p$ and $1/K_s$ are small relative to $1/K_L$. The only unknown, therefore, in Eq. (2) is the retardation time, τ_v. By fitting Eq. (2) to the measured retardation response, the term τ_v may be determined. The intrinsic permeability may then be expressed in dimensions of length squared as

$$k = \frac{\eta_L R^2}{\tau_v}\left(\frac{\beta b^2}{K_p} + \frac{\phi}{K_L} + \frac{b-\phi}{K_S}\right) \tag{4}$$

where R is the radius of the cylindrical specimen.

The measured K_s for 0.50 and 0.40 *w/c* cement pastes and a 0.50 *w/c* concrete are listed in Table 1. These values represent the average bulk modulus of the combined hydration products and unhydrated cement grains. Considering this, the values shown in Table 1 agree well with published nanoindentation results for measured elastic moduli of the hydration products and unhydrated cement grains [8-10]. Note that the values are relatively constant or slightly decreasing with age. This trend is sensible since the cement grains (high K_s) are gradually converted to hydration products (lower K_s).

Figure 3 shows the typical measured retardation response plotted in normal and log time. The response is fitted to Eq. (2) for determination of τ_v. The retardation response of materials between 0.4-0.6 *w/c* was measured, and the permeability calculated. The measured permeabilities, compared to permeabilities measured using the Beam Bending technique [11], are shown in Figure 4. Good agreement was found between the two techniques.

Table 1. Measured bulk modulus of the solid skeleton, K_s.

Strain x 10^{-6}	Pressure (Mpa)	Age (d)	K_s (GPa)	K_s (psi x 10^{-6})
0.50 *w/c* Paste				
16	2.52	7	53.15	7.71
39	4.41	9	37.71	5.47
47	6.90	14	48.90	7.09
63	6.90	17	36.48	5.29
0.40 *w/c* Paste				
Strain x 10^{-6}	Pressure (MPa)	Age (d)	K_s (GPa)	K_s (psi x 10^{-6})
11	2.06	5	62	8.99
37	2.52	17	23	3.34
0.50 *w/c* Concrete				
Strain x 10^{-6}	Pressure (MPa)	Age (d)	K_s (GPa)	K_s (psi x 10^{-6})
48	6.90	9	47.39	6.87
51	6.90	13	45.06	6.54

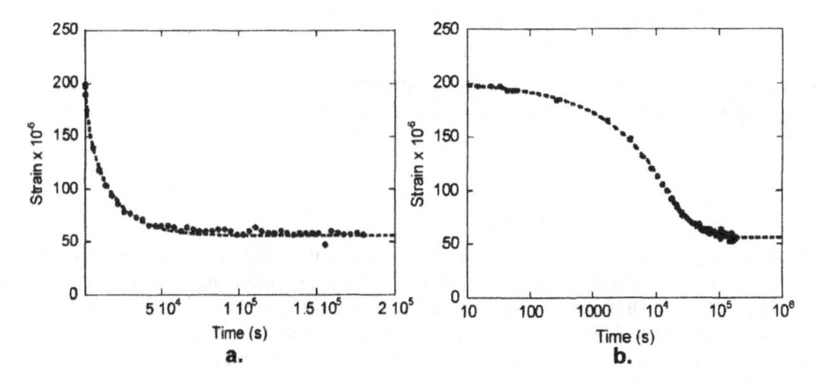

Figure 3. Typical retardation response fitted to Eq. (2). Compressive strain shown positive and on normal time (a.) and log time (b.) axes. Specimen is a 0.50 *w/c* paste subjected to 6.9 MPa hydrostatic pressure.

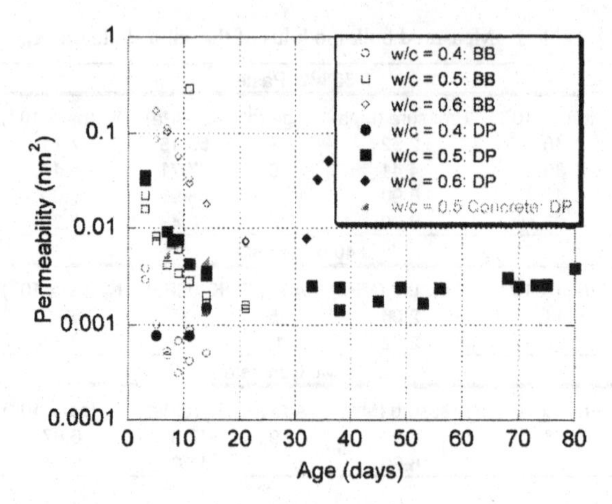

Figure 4. Comparison of measured permeability using DP test versus measured permeability using Beam Bending (BB) technique. BB data from [11].

ANALYZING TEST 2

When a sealed, partially saturated specimen is exposed to a hydrostatic pressure over a relatively short duration, the bulk modulus of the porous body may be determined as

$$K_p = \frac{p}{3\varepsilon_0}. \tag{5}$$

The analysis in [7] indicates that as little as 1% air in the pore fluid is all that is required for Eq. (5) to be appropriate since the pore fluid becomes highly compressible with even small amounts of air. A typical measured elastic response is shown in Figure 5. The response is quite linear in the range of applied hydrostatic pressures used in these tests. When the applied hydrostatic pressure is held constant with time, the sealed specimen will exhibit dilatational creep. Figure 6 shows the typical creep response of the sealed specimens.

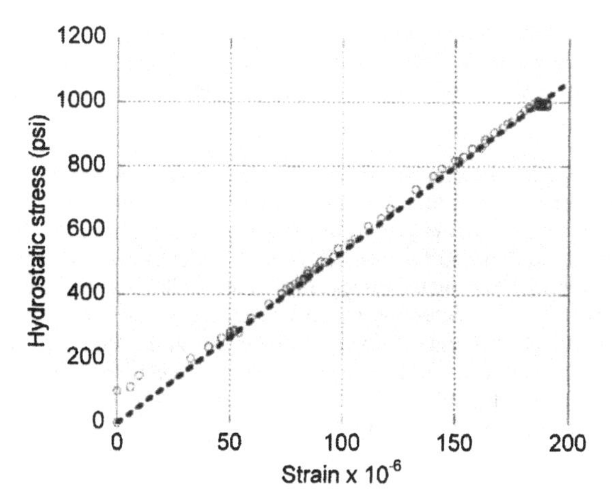

Figure 5. Measured elastic deformation response for Test #2. Specimen is 7 d old 0.35 Type I cement paste.

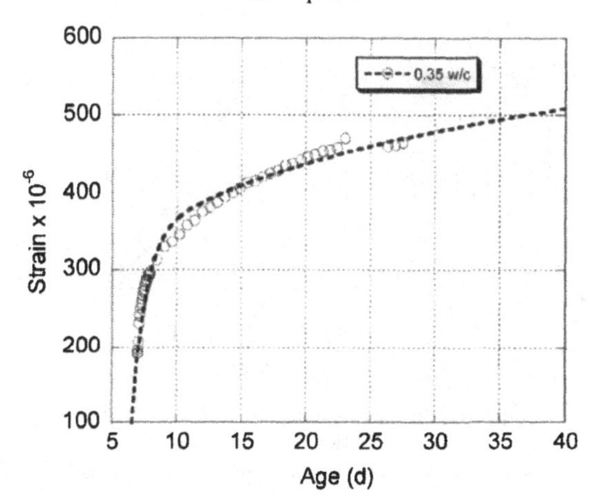

Figure 6. Measured dilatational creep under constant hydrostatic pressure. Autogenous shrinkage, measured on a companion specimen, was subtracted from measured creep response. Creep data shown fitted to aging Kelvin chain, where the aging is accounted for using Bazant's solidification theory [12]. Compressive strain is shown as positive.

CHALLENGES

There are a couple of challenges encountered in both Test 1 and Test 2. First, with Test 1, the presence of entrapped air voids dramatically alters the retardation response to applied hydrostatic pressure. Figure 7 demonstrates the effect of entrapped air on the measured axial strain. The entrapped air bubbles are gradually compressed and the retardation is not complete until the bubbles have fully diffused to the surface of the specimen. The effect of entrapped air has been described analytically by Scherer [7]. Another issue encountered in Test 1 is the potential to cause damage to the specimen, resulting in higher apparent permeability. This issue is discussed in detail elsewhere [6,7].

The greatest challenge in correctly performing Test 2 involves creating a seal that prevents fluid intrusion into the specimen even after long periods under hydrostatic pressure. This is necessary in order to measure the dilatational creep of the porous body, although it is not critical when simply obtaining the initial elastic response. When the seal is broken and some fluid intrudes into the partially saturated specimen, the specimen may exhibit a reduced creep rate or may re-expand slightly with time. The re-expansion occurs due to increases in pore fluid pressure similar to those occurring in Test 1. However, there is also dilatational creep occurring at the same time due to the complex state of stress in the material caused by partial saturation combined with the long test duration. Figure 8 illustrates an example of the measured response when the specimen is not properly sealed.

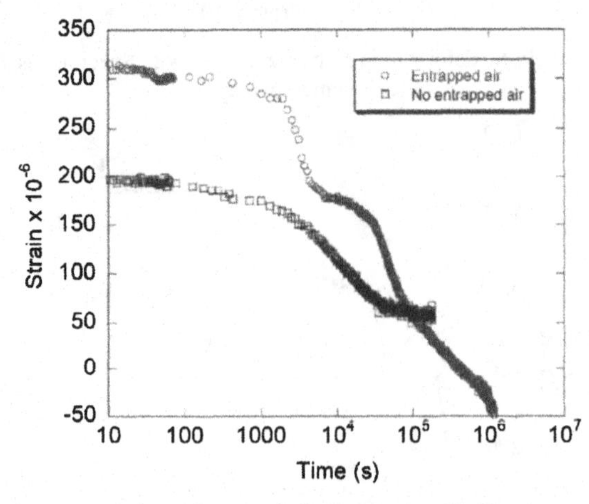

Figure 7. Effect of entrapped air on the retardation response of 0.50 *w/c* paste specimen. Note the inflection in the response of the specimen with entrapped air. The higher initial strain in the specimen with entrapped air is a result of the increase in compressibility of the pore fluid.

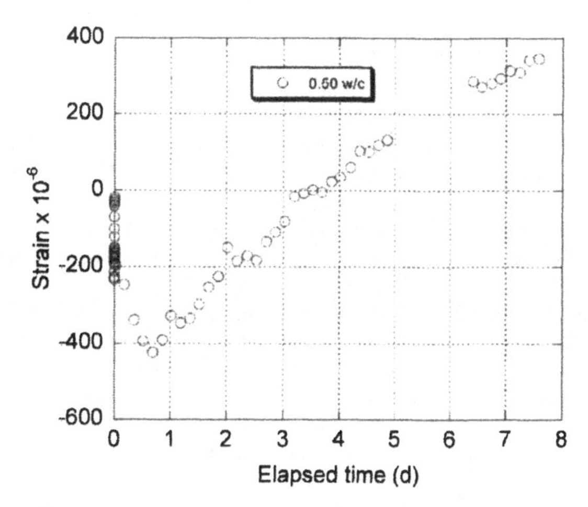

Figure 8. Example of creep specimen where seal leaked. Notice that the specimen creeps up to about 1 d after loading, then re-expands when fluid intrudes the specimen. Specimen is a mature 0.50 *w/c* cement paste. Compressive strain shown as negative.

SUMMARY

New experimental techniques and a new experimental apparatus have been described for measuring the permeability, k, the bulk modulus of the solid nanostructure, K_s, and the bulk modulus of the porous body, K_p. In addition, dilatational creep of the porous body may also be measured.

Measured permeability values agree well with permeabilities measured on identical materials using a different technique. The elimination of entrapped air is essential for obtaining quality results. The measured K_s values are reasonable when compared to nanoindentation results in the literature.

The new test method allows the measurement of dilatational creep. However, it is difficult to maintain a seal around the specimen for long periods of time, and leaks dramatically affect the measured results.

ACKNOWLEDGMENTS

Zachary Grasley was supported by an Eisenhower Fellowship through the National Highway Institute during this research. Additional laboratory support was provided by the Center of Excellence for Airport Technology.

REFERENCES

1. El-Dieb, A.S., Hooton, R.D., "Water-permeability measurement of high performance concrete using a high-pressure triaxial cell", *Cement & Concrete Research*, 1995, **25**(6): p. 1199-1208.
2. Grasley, Z.C., Lange, D.A., D'Ambrosia, M.D., *Drying Stresses and Internal Relative Humidity in Concrete*, in *Materials Science of Concrete VII*, J. Skalny, Editor. 2005, American Ceramic Society.
3. Bissonette, B., Marchand, J., Charron, J.P., Delagrave, A., Barcelo, L., "Early age behavior of cement-based materials", in *Mater. Sci. Concr. VI*, J. Skalny, Mindess, S., Editor. 2001.
4. Bentz, D.P., E.J. Garboczi, and D.A. Quenard, "Modelling drying shrinkage in reconstructed porous materials: application to porous Vycor glass", *Modelling and Simulation in Materials Science and Engineering*, 1998, **6**(3): p. 211-36.
5. Lura, P., Guang, Y.E., van Bruegel, K., "Effect of cement type on autoegenous deformation of cement-based materials", *ACI*, 2004, **SP 220-4**: p. 57-68.
6. Grasley, Z. C., Scherer, G. W., Lange, D. A., and Valenza II, J., "Dynamic pressurization method for measuring permeability and modulus: II. Cementitious materials", *Materials & Structures,* accepted for publication, 2006.
7. Scherer, G. W., "Dynamic pressurization method for measuring permeability and modulus: I. Theory", *Materials & Structures,* accepted for publication, 2006.
8. Velez, K., Maximilien, S., Damidot, D., Fantozzi, G., and Sorrentino, F., "Determination by nanoindentation of elastic modulus and hardness of pure constituents of Portland cement clinker." *Cement and Concrete Research*, 2001, **31**(4): p. 555-561.
9. Acker, P. "Micromechanical analysis of creep and shrinkage mechanisms." *ConCreep 6*, MIT, 2001.
10. Constantinides, G., and Ulm, F.-J., "The effect of two types of C-S-H on the elasticity of cement-based materials: Results from nanoindentation and micromechanical modeling." *Cement and Concrete Research*, 2004, **34**(1): p. 67-80.
11. Vichit-Vadakan, W., Scherer, G. W.,"Measuring permeability of rigid materials by a beam-bending method: III, cement paste." *Journal of the American Ceramic Society*, 2002, **85**(6): p. 1537-1544.
12. Bazant, Z. P., "Viscoelasticity of solidifying porous material - Concrete", *J. of the Eng. Mech. Div., ASCE*, 1977, **103**(EM6): p. 1049-1067.

USING MONTE CARLO ANALYSIS TO QUANTIFY HOW THE NUMBER AND ARRANGEMENT OF SENSORS INFLUENCE THE ACCURACY OF AN IN SITU SENSING SYSTEM

Tom Schmit
Purdue University, School of Civil Engineering
550 Stadium Mall Drive
West Lafayette, Indiana, 47907-1284, USA

Jason Weiss
Purdue University, School of Civil Engineering
550 Stadium Mall Drive
West Lafayette, Indiana, 47907-1284, USA

ABSTRACT

The use of in situ sensing systems provides the potential to accurately determine the properties of in place concrete. These properties can be used as input parameters for service-life prediction models for the purpose of determining the projected service life of a given facility [1,2,3]. The accuracy and economic viability of in situ sensing technologies is dependant upon both the precision and number of sensors utilized in the sensing system as well as the arrangement of the sensors. In situ measurements should not be used to simply to perform as many measurements of the structure as possible, but rather to simply obtain the amount of information necessary to accurately simulate the behavior of the in place concrete. Designing in situ sensing systems in this manner insures an efficient application of sensing technologies. This paper describes a methodology to allow a sensing system designer to quantify how the number and arrangement of sensors influences the variability of the information obtained from a sensing system. This information can provide a sensing system designer with the ability to make informed decisions as to the number of sensors required for a particular sensing application as well as the best way to arrange these sensors inside the concrete.

THE POTENTIAL FOR IN SITU SENSING

Engineers are frequently faced with several design options that may improve the durability of a concrete structure. The selection of the best design option cannot simply be based on engineering intuition alone. To aid engineers in making these design decisions, computational service-life prediction models have been developed which simulate the impact of different design options on the long-term costs of the concrete facility. Commercially available software packages [3,4,5,6,7] have been developed which are capable of performing service-life simulations and the accompanying economic analysis.

In addition to aiding engineers in the design process, investigations have shown that performance prediction models can be used for determining payment adjustments for performance related specifications (PRS). In one such application, payment adjustments were made for a portland cement concrete pavement based on a life-cycle cost analysis calculated by a performance prediction model [2].

Irrespective of whether these models are used for making decisions pertaining to the economic benefits of different design options or for determining payment adjustments for performance related specifications, they require material property inputs (e.g., strength, elastic modulus, diffusion coefficient). Accurately simulating the life-cycle performance of in place concrete requires the precise determination of the input material properties. Traditionally, these material properties have been obtained from tests performed on companion specimens or are approximated using empirical relations [5,7]. However, due to differences in sampling, consolidation, and curing, the properties of companion specimens may differ from the properties of the concrete placed in structures [8]. Furthermore, empirical relations have the potential to predict concrete properties that do not always describe those of an in place concrete. Thus, utilizing companion specimens and empirical relations to provide material property inputs for models can result in inaccurate simulations and predictions [9]. In situ measurements may provide an alternate method for obtaining material property inputs for computational performance models [1]. Unlike companion specimens and empirical estimates, in situ measurements provide information on the properties of the in place concrete. The inputs obtained from in situ measurements may, therefore, more accurately reflect the material that is in service and could potentially lead to more accurate simulations.

THE ADVANTAGES OF EFFICIENT IN SITU SENSING SYSTEMS

From an economic standpoint, the cost of a measurement is typically dependent upon the level of precision that is required from that measurement. This is because increasingly precise measurements frequently require more complex and costly instruments. Some sensing systems employ multiple sensors. In these sensing applications the economy of the system is not only dependent on the precision of the sensors, but it is also dependent on the number of sensors utilized in the system. The hardware and software required to record and transmit the data generated by in situ sensors can be very costly. Limiting the number of sensors employed in an in situ sensing system not only reduces sensor costs, but also reduces the costs associated with recording and transmitting the data generated by the sensors. Over-designing sensing systems by either designing unnecessarily precise sensors or by employing more sensors than necessary is not an economically sound practice.

Economic considerations do not provide the only motivation for limiting the number of sensors utilized in a sensing system. In some systems, including cementitious systems, the presence of the sensor may actually alter the system undergoing measurement. Sensors in concrete can introduce transition zones between the cement paste and the sensors that would not normally be present. Additionally, sensors can introduce stresses that may cause the concrete to crack around the sensor [10]. Finally, the introduction of sensors can interfere with the normal consolidation process by altering the distribution of aggregates and paste.

RESEARCH OBJECTIVE

The objective of this work is to develop a methodology that may be applicable to many sensing applications that will allow a sensing system designer to quantify how the number and arrangement of in situ sensors affects the variability of the information obtained from a sensing system. This information can provide a sensing system designer with the ability to make informed decisions as to the number of sensors required for a particular sensing application as well as the best way to arrange these sensors inside the concrete. In addition, this information will provide a means to determine whether a system is over-designed and will help prevent problems associated with over-designed sensing systems discussed in the previous section.

OVERVIEW OF THE RESEARCH APPROACH

A hypothetical chloride sensing system was used to develop the Monte Carlo methodology. The system consists of hypothetical chloride sensors placed in a concrete at various depths from the surface. These sensors are capable of measuring the concentration of chlorides in the surrounding concrete. The sensors are assumed to have an appropriate resolution and a negligible drift over a wide range of time and temperature. Finally, the sensors are assumed to be durable enough to perform chloride measurements throughout the entire service life of the structure. Future work is aimed at quantifying the influence of variations that may be caused by sensor drift or individual sensor failure.

The system undergoing measurement in this work is a simulated reinforced concrete element that is exposed to chlorides from one surface. Effort was taken to assign properties to this simulated concrete that are consistent with the properties observed for in-service reinforced concretes. The 'actual values' of the simulated concrete are defined by the investigator and, therefore, can be considered to be known exactly. This justifies the comparison of a 'simulated measurement' of a property of the simulated concrete to the 'actual value' (i.e., the 'actual value' is the exact value) of that property.

In order to obtain a somewhat realistic distribution of chlorides in the simulated concrete, the software LIFE-365™ [5] was used to generate a chloride profile. LIFE-365™ enabled the effects of increased hydration, temperature variation, and changes in the surface chloride concentration to be considered in the assigned 'actual value' for the chloride distribution inside the simulated reinforced concrete. The LIFE-365™ simulation was based on a concrete that was in service in a bridge deck in Indianapolis, Indiana for 10 years. The profile generated by LIFE-365™ is shown in Figure 1.

The analysis performed later in the paper requires an analytical solution for the 'actual values' of the chloride distribution. Therefore the data from the Life-365™ simulation was fitted with an analytical solution to Fick's second law (Equation 1)

$$C(x,t) = C_s \cdot \left(1 - \mathrm{erf}\left[\frac{x}{2\sqrt{D \cdot t}}\right]\right) \tag{1}$$

where $C(x,t)$ is the concentration of ions, x is depth measured from the surface, t is time, D is the diffusion coefficient, and C_s is the concentration of ions at the surface [11].

The solution to Fick's second law (Equation 1) was fit to the data generated by the LIFE-365™ simulation (shown in Figure 1) using least squares regression, and the fit resulted in a diffusion coefficient of 1.53×10^{-12} m^2/s and a surface chloride concentration of 7.77 kg/m^3. Throughout the remainder of this work, the fit of Fick's second law will be referred to 'actual chloride profile', and simulated measurements will be performed on this actual chloride profile.

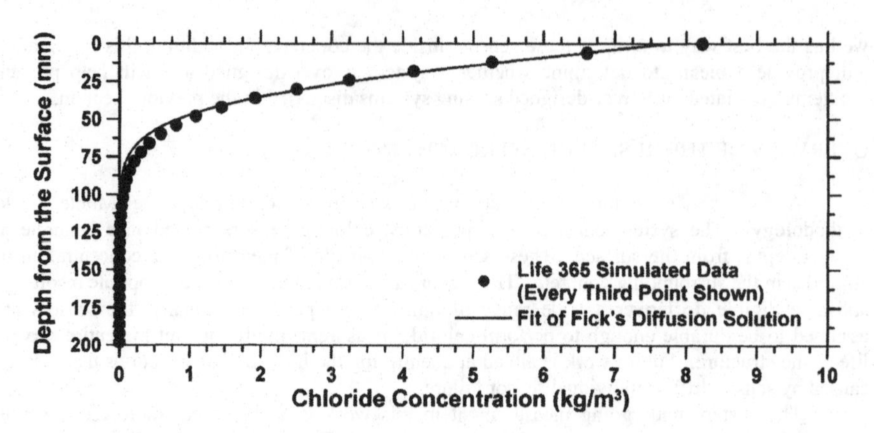

Figure 1: The best fit of the solution of Fick's second law to the chloride profile generated by LIFE-365™ software

Although diffusion is only one of several mechanisms responsible for chloride ingress in non-saturated concretes, the simplicity of Equation 1 made it a logical starting point for the initial investigation. The techniques described in this paper can be modified to account for other transport mechanisms. An added benefit of using Equation 1 is the fact that a similar solution has been used to describe the relative humidity profile in a drying concrete specimen [12] and the temperature distributions in mass concrete [13]. It is the investigators hope that minimal modification are required to allow the current work to be applied to relative humidity and temperature sensing systems.

SOURCES OF VARIABILITY

There are four inherent sources of variability that can be considered in Fick's second law. The following section briefly discusses these sources of variability, and the reader is referred to reference [14] for a more detailed discussion.

The first source of variability is associated with the value of the concentration, $C(x,t)$, measured by the sensor. It is assumed that this variability is normally distributed and centered on the actual value of the chloride concentration. In order to select a value for this variability, a range of possible values was first estimated.

ASTM C 1152, The Standard Test Method for Acid Soluble Chloride in Mortar and Concrete [15], was used to select the lower endpoint for the variability in the chloride concentration measurement. The ASTM standard states the multilaboratory standard deviation of the test method as 0.0475 kg/m^3. This standard deviation was achieved in the laboratory in a carefully controlled environment. Therefore, this standard deviation was taken as the lower limit of a sensing system performing in situ concentration measurements in a much less controlled environment.

The upper endpoint for the variability in measuring the chloride concentration was selected based on the threshold chloride concentration. This concentration falls somewhere in the range of 0.9 to 1.2 kg/m^3 [16]. For this work, a value of 1.0 kg/m^3 is assumed to be the threshold chloride concentration. If the magnitude of the variability of a chloride sensor is near the magnitude of the threshold concentration, the measurements obtained from the sensor can not provide information about when this critical concentration has been exceeded. A measurement variability equal to 25% of the threshold concentration was deemed the largest acceptable practical level of variability for a chloride sensor. A sensor with this variability would still allow information about the threshold chloride level to be obtained. The probability that a measurement falls within two standard deviations of the mean is 95%. Consequently, the upper endpoint for the standard deviation in the variability in measuring the chloride concentration was selected to be 0.125 kg/m^3. A standard deviation of 0.125 kg/m^3 results in 95% of the measurements obtained by a sensor measuring a chloride concentration equal to the threshold concentration to be within 25% of the threshold concentration.

The lower and upper endpoints for the variability in the chloride concentration measurements performed by a chloride sensor were estimated to be 0.0475 kg/m^3 and 0.125 kg/m^3, respectively. The midpoint of this range, or a standard deviation of 0.086 kg/m^3, was used in the simulations performed in this work.

The second source of variability in Fick's second law is related to the location, x, of the sensor relative to the concrete surface. This variability results from the fact that it is difficult to position a sensor exactly at the desired distance from the concrete surface. The current work assumes that this variability is normally distributed and centered about the depth where the system designer intends for the sensor to be located. In order to select a value for this variability, a range of possible values was first estimated.

The lower endpoint for the variability in locating a sensor in concrete is based on the method of placing the sensor by attaching it to the end of a rigid object and pushing the sensor to the desired distance from the surface of the plastic concrete. It is estimated that placing a sensor using this technique could result in the sensor being at most 5 mm from the desired location. This resulted in selecting the lower endpoint standard deviation to be 2.5 mm which assures that 95% of the time the sensor will be within ±5 mm of the desired location.

The upper endpoint for the variability in locating a sensor in concrete is based on the method of positioning the sensor relative to the steel reinforcement before the concrete is placed. In this case the variability in the distance the sensor is placed from the surface of the concrete is equal to the variability in the thickness of the concrete cover placed over the reinforcement. A study by Van Deveer [17] on 62 bridges in 17 states observed that the standard deviation of the concrete cover on the bridge decks investigated was 9.5 mm (3/8 inches). Consequently, the upper endpoint for the variability in positioning a sensor was estimated to be 9.5 mm.

The lower and upper endpoints for the variability in positioning a chloride sensor in concrete were estimated to be 2.5 mm and 9.5 mm, respectively. The midpoint of this range, or a standard deviation of 6.3 mm, was used in the simulations performed in this work.

The third source of variability in Fick's second law is the surface concentration, C_s. The surface concentration varies from structure to structure as well as from location to location on the same structure (or even from day to day at the same location in a structure). Consequently, this quantity has a large associated variability. However, when multiple sensors are used to measure a chloride profile the surface concentration can be determined by fitting Equation 1 to the concentration measurements using least squares regression. This technique is employed in the current work; and as a result, the variability in the surface concentration is not explicitly considered.

The fourth source of variability in Fick's second law is the time, *t*. It is assumed that time is known to a reasonable degree of accuracy; and thus the variability in the time is negligible.

Table 1 provides a summary of the standard deviations that will be assigned to the sources of variability considered in the simulations described in this paper.

Table 1: The values assigned to the variability in the chloride sensing system

Quantity Contributing the Variability	Estimated Standard Deviation	Units
Depth of Sensor from Surface, x	6.3	mm
Concentration Measurement C(x,t)	0.086	kg/m^3

THE MONTE CARLO SIMULATION PROCEDURE

The Monte Carlo technique described in this section simulates field measurements performed on the chloride profile defined in Section 4.0. Figure 2 provides a schematic representation of how the Monte Carlo simulations are performed.

The simulations begin by defining the 'desired sensor locations'. The desired sensor locations are the exact locations where the system designer expects the sensors to be placed inside the concrete. However, in the field, the sensors will not be placed exactly at the desired locations, and the variability involved in positioning the sensors is simulated by applying a normal distribution to the desired sensor locations. The normal distribution is centered about the desired location, and the standard deviation is equal to the value provided for this variability in Table 1. The locations in the concrete where the sensors are positioned after the placement variation is applied to the desired sensors locations are deemed the 'actual sensor location'. The process of applying variability to the desired sensor locations is shown in Step 2 of Figure 2.

With the actual sensor locations determined, the chloride concentrations are calculated at these actual locations using Equation 1. Once the chloride concentrations at the location of the sensors are calculated, the variability in the concentration measurements is simulated by applying a normal distribution to the calculated concentrations (Step 3 in Figure 2). The normal distribution is centered about the calculated chloride concentration value and the standard deviation is equal to the value provided for this variability in Table 1.

Rather than being positioned at the exact location desired by the system designer, the sensors in the field are misplaced. The system designer, however, has no knowledge as to the actual locations of the sensors inside the concrete and incorrectly assumes the sensors are positioned at the desired locations. The result is that the concentration measurements obtained at the actual sensor locations are treated by the system designer as though they were obtained at the desired sensor locations. To simulate this, the Monte Carlo simulation attributes the measured chloride concentrations to the desired sensor locations despite the fact that they were obtained at the actual sensor locations (Step 4 in Figure 2).

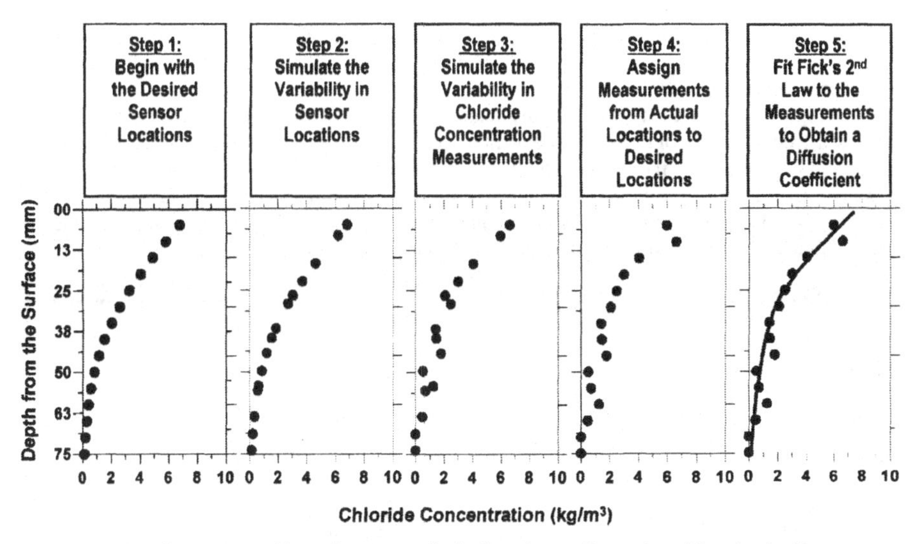

Figure 2: An Illustration of how the Monte Carlo Simulation Procedure Was Applied

Simulating the variability in sensor placement and chloride concentration measurement with normal distributions creates a problem for sensors located near the concrete surface and concentration measurements near zero. The problem arises when the statistical variation in these quantities results in a sensor being placed at a negative depth (above the surface of the concrete) or a sensor measuring a negative chloride concentration. Not only are both of these situations physically impossible, the error function in Equation 1 is not defined for negative concentration and depth values. The simulations account for this problem by automatically correcting for negative values. Any sensor position that is assigned a negative depth is automatically reassigned to the surface at depth of zero. Similarly, negative concentrations are corrected by reassigning a value of zero to the chloride concentration measurement.

The data obtained from the Monte Carlo simulations consists of simulated chloride concentration measurements and the corresponding intended depths to which the concentration measurements are attributed. The final step of the simulation is to fit the solution of Fick's second law to the data (Step 5 in Figure 2). The diffusion coefficient and the surface concentration in Equation 1 are optimized to obtain the best fit to the simulated data. The optimized value of the diffusion coefficient for the fit is then recorded and output from the Monte Carlo simulation.

It should be mentioned that because a fit is used to determine the diffusion coefficient of the concrete from the chloride concentration measurements, the Monte Carlo simulations can only be performed for systems with three or more sensors. This is because three data points are needed in order to produce a fit of the solution to Fick's second law.

Before the Monte Carlo methodology could be used to investigate sensing systems, it was necessary to determine the appropriate number of iterations necessary to obtain reproducible results. It was determined that the values of several separate simulations converged between

3,000 and 10,000 iterations. As a result, the simulations discussed in this paper were performed using 10,000 iterations. The reader is referred to [14] for more details on how the number of iterations was determined.

Figure 3 shows the typical outputs obtained from the Monte Carlo simulations. Figure 3a shows a frequency histogram with the values for the diffusion coefficients obtained from the simulations plotted on the x-axis and the number simulations in which these values were obtained (per 10,000 simulations) plotted on the y-axis. Figure 3b is a cumulative distribution function which shows the frequency of simulation results on the (y-axis) in which the observed value of the diffusion coefficient was larger than a particular value on the x-axis.

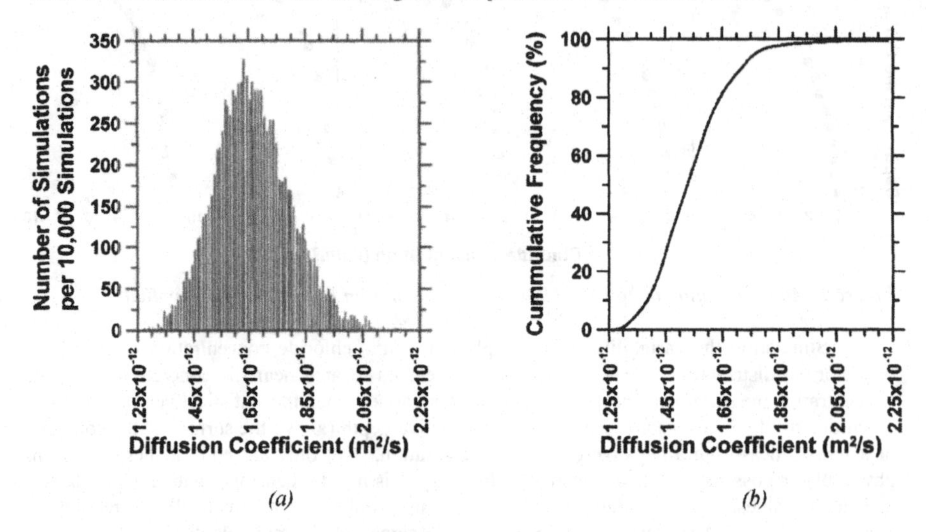

Figure 3: Typical Outputs of the Monte Carlo Simulations

The Monte Carlo output shown in Figure 3 can be used to determine the variability in the time to corrosion initiation that would be predicted from a life-cycle simulation. This is achieved by utilizing statistical software capable of providing the range of values of diffusion coefficients for a selected confidence interval. In this study a 95% confidence interval was chosen. The 95% confidence interval is interpreted as meaning that 95% of the values of the diffusion coefficients obtained by the Monte Carlo simulation fall within the interval.

In order to convert from a diffusion coefficient value to the number of years until the reinforcement begins to corrode, the software package LIFE-365™ was utilized. The diffusion coefficients are entered into the same LIFE-365 ™ scenario discussed in Section 4.0, and the software performs the calculations to determine the time to corrosion initiation. Analyzing the upper and lower bounds of the 95% confidence interval of diffusion coefficient values in LIFE-365 ™ converts the values to a 95% confidence interval for time to corrosion initiation. Thus, the effects of the number and arrangement of sensors can be ascertained by comparing the sizes of the 95% confidence intervals for time to corrosion initiation for each sensor arrangement.

MONTE CARLO SIMULATION RESULTS

The first set of experiments performed using the Monte Carlo simulations investigated how the number of sensors utilized in a sensing system affects the predicted time to corrosion initiation. In the experiments, the sensors were equally spaced throughout the concrete cover. The results of the experiments are provided in Figure 4 and indicate that as the numbers of sensors increases the variability in time to corrosion initiation decreases. It can be observed that when the sensors are equally spaced throughout the concrete cover, the variability decreases in proportion with the square root of the number of sensors.

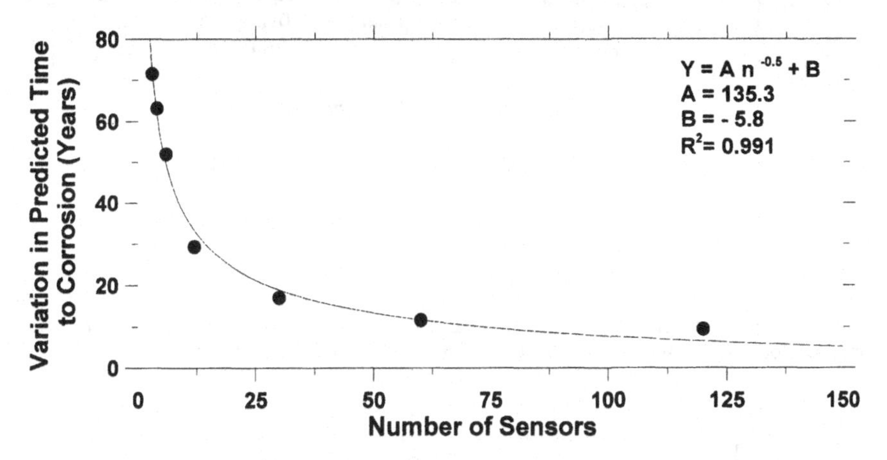

Figure 4: Results of the Monte Carlo simulations performed to quantify the effect of the number of equally spaced sensors on the variability of the sensing system.

The second set of experiments performed using the Monte Carlo simulations investigated how the spatial arrangement of sensors affects the variability of the sensing system. Three sensor configurations were investigated for the second set of experiments, and each of these configurations had one sensor located at the depth of the reinforcement.

The first sensor configuration investigated is referred to as the 'equally spaced sensor arrangement'. In this arrangement the sensors are equally spaced throughout the concrete cover. The second sensor configuration is referred to as the 'surface sensor arrangement'. This sensor arrangement is the same as the equally spaced sensor arrangement with the addition of one sensor located 1mm from the surface. The final sensor configuration investigated is referred to as the 'square root of depth sensor arrangement'. In this arrangement, the distance between consecutive sensors is squared. Figure 5 provides an example of each of the sensor arrangements.

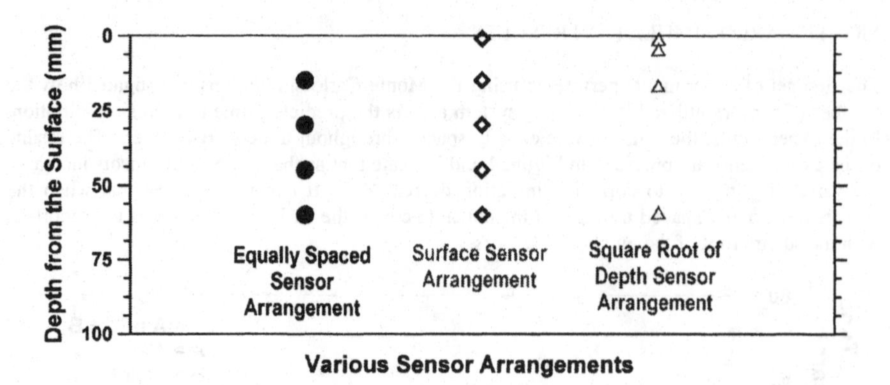

Figure 5: A graphical representation of the three sensor arrangements investigated

The results of the Monte Carlo simulations investigating the spatial arrangement of sensors are provided in Figure 6. The performance of all three sensor arrangements converge as the number of sensors increases. This is expected due to that fact that the differences between the three arrangements are very minor in systems containing high numbers of sensors. In Figure 6, the difference in performance of all three sensor arrangements is negligible for systems containing 60 and 120 sensors and the data points for each arrangement overlap.

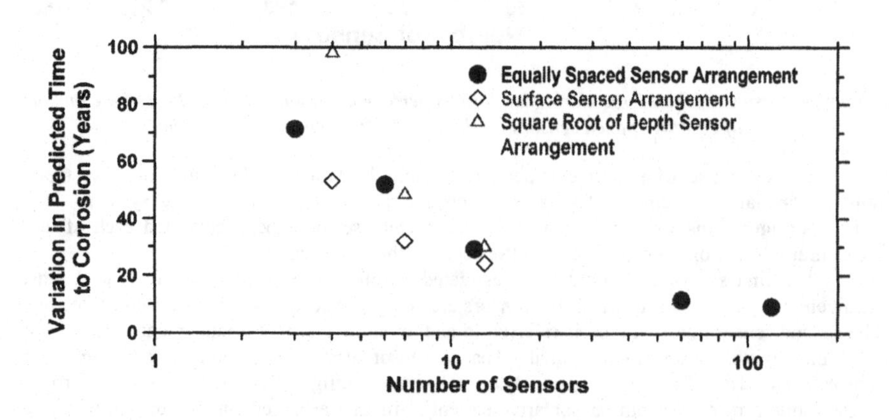

Figure 6: The results of the Monte Carlo simulations investigating the how the arrangement of sensors affects the variability of a sensing system.

A comparison between the performance of the equally spaced sensor arrangement and the surface sensor arrangement in Figure 6 reveals that for sensing systems containing less than 15 sensors the addition of a surface sensor located 1.0 mm from the concrete surface significantly decreases the variability in the predicted the time until corrosion initiation. For example, the sensing configuration utilizing three equally spaced sensors has a variability of 71.5 years, while the same sensor configuration with the added surface sensor has a variability of 53.2 years. The one additional sensor placed 1.0 mm from the concrete surface reduced the variability by 18.3 years. Realistically, both of these sensing configurations are inadequate due to their enormous variability, but are useful for highlighting the benefit of a surface sensor. The large variability in the time to corrosion initiation for these simulations stems not only from the low number of sensors performing measurements, but also stems from the variability assigned to concentration measurements for these simulations (Table 1). When additional simulations [provided in ref 14] were performed on the same sensor configurations using a smaller value for the variability in the chloride concentration measurements, the sensor configurations had significantly smaller confidence interval widths.

The results in Figure 6 also allow a quantitative comparison to be made between the performance of the square root of depth spacing arrangement and the other two sensor arrangements. The results in Figure 6 indicate that for sensor arrangements containing less than about 20 sensors the square root of depth sensor arrangement is the poorest performing arrangement. This poor performance may be due to the fact that the square root of depth sensor arrangement over-samples the concrete near the surface. Small variations in the concentration of chlorides at depths near the rebar make a large difference in the predicted diffusion coefficient. Possessing more measurements at the surface means when a least squares regression is performed the best fit will over weight the surface measurements and under weight the measurements near the reinforcement. This under weighting of the measurements near the reinforcement results in a large variability in the predicted the time until corrosion initiation.

CONCLUSIONS

The goal of this research was to demonstrate a methodology that would be capable of insuring the efficient application of sensing technologies by preventing the over sampling of a concrete. Toward this end, a Monte Carlo analysis technique was developed based on a chloride sensing system; however, the technique can be adapted to a multitude of sensing applications.

The first round of Monte Carlo simulations performed in this work investigated how the number of equally spaced sensors employed in a chloride sensing system affects the variability in the predicted time to corrosion initiation. The results of the simulations were provided in Figure 4 and indicated that the variability in the predicted time to corrosion initiation is inversely proportional to the square root of the number of sensors employed in the sensing system.

The second round of Monte Carlo simulations performed in this work investigated how the arrangement of the sensors employed in the chloride sensing system affects the variability in the predicted time to corrosion initiation. Three sensor arrangements were investigated: the 'equally spaced' sensor arrangement, the 'surface sensor' arrangement, and the 'square root of depth' sensor arrangement.

The results of the second round of simulations indicated that the performance of all three sensor arrangements converge for systems employing over 15 sensors. This was due to the fact that the differences between the three arrangements are negligible for systems with high sensor numbers.

The simulations determined that the surface sensor arrangement was the most efficient sensor arrangement investigated, and the poorest performing sensing arrangement investigated was the square root of depth sensor arrangement. A comparison between the surface sensor arrangement and the equally spaced sensor arrangement lead to the conclusion that the addition of one sensor near the surface results in considerable reduction in the variation in the predicted time to corrosion for sensing systems with 15 or less sensors.

The Monte Carlo methodology developed in this work can be a valuable tool for an in situ sensing system designer. The results in this paper have shown that the Monte Carlo method can be used to aid in determining the variability that can be expected for a particular sensing system. Additionally, the technique can provide a sensing system designer with the ability to make informed decisions as to the number of sensors required for a particular sensing application as well as the best way to arrange these sensors inside the concrete. Possessing this information allows a sensing system designer to avoid the inefficient over-sampling of a concrete.

ACKNOWLEDGEMENTS

The authors gratefully acknowledge support received from the National Science Foundation (NSF) through Grant No. 0134272. Any opinions, findings and conclusions or recommendations expressed in this material are those of the authors and do not necessarily reflect the views of the National Science Foundation (NSF).

REFERENCES

1. Weiss, W.J. (2001) "Linking Insitu Monitoring with Damage Modeling for Life-Cycle Performance Simulations of the Concrete Infrastructure," NSF Career Development Plan, National Science Foundation

2. Graveen, C., Weiss, W. J., Olek, J., Nantung, T., and Gallivan, V. L. (2004) "The Implementation of a Performance Related Specification (PRS) for a Concrete Pavement in Indiana," Submitted to the Transportation Research Board

3. Hoerner, T. E., and Darter, M. I. (2000) "Improved Prediction Models for PCC Pavement Performance-Related Specifications, Volume II: PaveSpec 3.0 User's Guide," FHWA RD-00-131, Federal Highway Administration, Washington, D. C.

4. Ehlen, M. A. (1999) "BridgeLCC 1.0 Users Manual: Life-Cycle Costing Software for Preliminary Bridge Design," National Institute of Standards and Technology, Gaithersburg, Maryland

5. Thomas, M. D. A., and Bentz, E. C. (2001) "LIFE-365™ Service Life Prediction Model, Users Manual," Silica Fume Association

6. Marchand, J. (2001) "Modeling the Behavior of Unsaturated Cement Systems Exposed to Aggressive Chemical Environments," Materials and Structures, Vol. 34, pp. 195 – 200

7. Ishida, T., and Maekawa, K. (1999) "An Integrated Computational System for Mass/Energy Generation, Transport, and Mechanics of Materials and Structures," Translation from Proceedings of JSC, Vol. 44, pp. 627

8. Skalny, J., and Idorn, G. M. (2004) "Thoughts on Concrete Durability," International Symposium of Advances in Concrete through Science and Engineering: Proceedings of the International RILEM Symposium, Evanston, Illinois

9. Hooton, D., Thomas, M.D.A., Marchand, J., and Beaudoin, J. (2001) "Ion and Mass Transport in Cement-Based Materials," Materials Science of Concrete Special Volume, The American Ceramics Society, Westerville, Ohio

10. Weiss, W.J., Shane, J.D., Meises, A., Mason T.O., and Shah, S.P. (1999) "Aspects of Monitoring Moisture Changes Using Electrical Impedance Spectroscopy," Self-Desication and its Importance in Concrete Technology, Proceedings of the Second International Research Seminar in Lund, June 18, 1999, pp. 31-48

11. Crank, J. (1956) "The Mathematics of Diffusion," Oxford University Press, London, England

12. Moon, J. H., Rajabipour, F., and Weiss, J. (2004) "Incorporating Moisture Diffusion in the Analysis of the Restrained Ring Test," The 4th International Conference on Concrete under Severe Conditions of Environment and Loading (CONSEC'04), Seoul, Korea. pp. 1973-1980

13. Mehta, P.K., and Monteiro, P. (1993) "Concrete: Structure, Properties, and Materials," 2nd Edition, Prentice Hall, New Jersey, NY

14. Schmit, T. (2005) "A Fundamental Investigation on Utilizing Error Propagation, Monte Carlo Simulation, and Measurement Interpretation Techniques to Design Efficient In Situ Covercrete Sensing Systems," Purdue University, August 2005.

15. ASTM C 1152/C 1152M – 04 "Standard Test Method for Acid-Soluble Chloride in Mortar and Concrete," ASTM International, West Conshohocken, PA

16. Bentur, A., Diamond, S., and Berke, N.S. (1997) "Steel Corrosion in Concrete: Fundamentals and Civil Engineering Practice," E & FN Spon, London, UK

17. Van Daveer, J.R. (1975) "Techniques for Evaluating Reinforced Concrete Bridge Decks," ACI Journal, December, pp. 697-704.

ROLE OF *IN SITU* TESTING AND MONITORING IN ASSESSING THE DURABILITY OF REINFORCED CONCRETE

P.A.Muhammed Basheer[1], Kenneth V. Grattan[2], Tong Sun[2],
W. John McCarter[3], Adrian E. Long[1], Daniel McPolin[1], Lulu Basheer[1]

[1]Queen's University Belfast, School of Civil Engineering, David Keir Building, Stranmillis Road, Belfast, Northern Ireland, UK, BT9 5AG;
[2]City University, School of Engineering and Mathematical Sciences, London, UK, EC1V 0HB;
[3]Heriot-Watt University, School of Built Environment, Edinburgh, UK, EH14 4AS.

ABSTRACT

Durability of concrete is related to its transport properties, such as absorption, diffusion and permeability. It has been possible to relate the transport properties to a range of durability parameters in accelerated laboratory tests. However, there is a need to explore how these relationships can be used for the service life prediction of reinforced concrete structures. *In situ* tests can provide useful and extremely important data, but sensors embedded at the time of construction could provide data on a continual basis, which also eliminates the need to carry out regular investigations on the structure to assess its condition. In the context of the durability of reinforced concrete structures, it is the pH, ionic content and moisture content are three parameters which are extremely important to monitor. Developments in electrical based and fibre optic based sensors can be used in developing concrete sensors for implanting and continuous monitoring of structures.

In this paper, progress made at Queen's University Belfast in using two in-situ permeability apparatus, viz. Autoclam Permeability System and Permit Ion Migration Test, for the measurement of the gas/water permeability, water absorption and the ion diffusion through the concrete cover is summarised. In addition, fibre optic sensor systems are discussed for installation and long-term monitoring of the pH, moisture and chloride content in concrete. An electrode array system is described, which is capable of monitoring the changes to the electrical resistance of concrete as a result of the chloride ingress and similar other effects. Some typical results are presented for these novel monitoring and testing systems to illustrate their usefulness for the condition monitoring and assessment of reinforced concrete structures.

INTRODUCTION

Significance of Testing and Monitoring Structures

A survey of most frequently reported cases of deterioration of reinforced concrete structures has indicated that the deterioration of reinforced concrete structures is primarily caused by the corrosion of the embedded steel (Fig. 1) when a chloride front or a carbonation front reaches the reinforcement. The current situation in most developed countries is that repair and rehabilitation costs of structures far exceed the total budget for capital development programmes. The issue with most structures is not *if* maintenance is required, but *when* – and when to schedule it most cost-effectively. Smart structures – structures incorporating in their design sensor elements and actuating devices which not only diagnose a problem but effect a solution to the problem – represent one approach for the diagnosis of faults and incipient failures and to schedule maintenance effort most effectively and at optimum cost, minimising the possibility of catastrophic failure. Effective monitoring requires a knowledge of physical parameters, such as temperature, strain and pressure and chemical parameters, such as pH and chloride content. Fig. 2 illustrates the usefulness of monitoring at different stages of the performance of a structure. Ideally information collected before the extent of the problem becomes severe, i.e. in the initiation period, will be invaluable for the effective maintenance management of civil engineering structures.

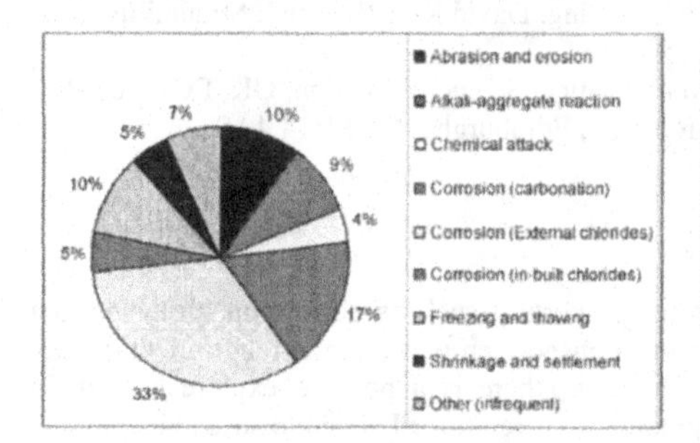

Fig. 1 Frequency of reported failure mechanisms in reinforced concrete structures

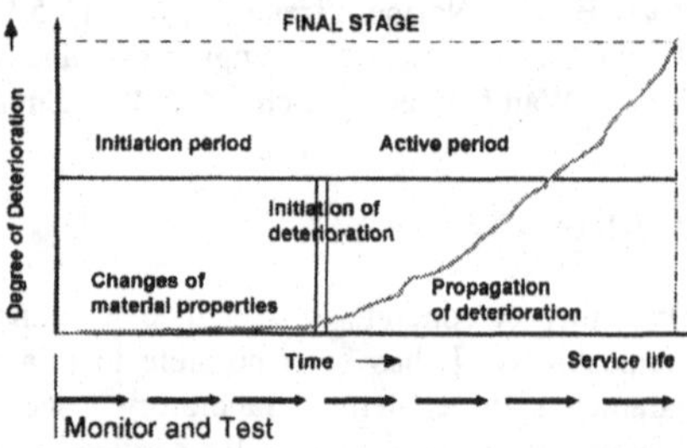

Fig. 2 Usefulness of continuous monitoring of structures

Figure 3 illustrates that the deterioration of reinforced concrete is related to its microstructure and the transport of the aggressive substances. The permeation properties of concrete are related also to its microstructure and the degree of saturation. Thus an assessment of the durability of concrete structures can be made in terms of the measured permeation properties.

Reinforced concrete structures can be made durable by resorting to appropriate design, selection and use of materials and construction practices. If an extremely high initial quality is achieved, this may lead to a performance history as shown by curve 1 in Fig. 4. However, for most concretes, the performance history is such that the performance deteriorates gradually with time and not suddenly (and catastrophically) as shown by curve 2, but there is significant deterioration with time, and intermittent maintenance changes the performance level or alters the rate of change of the performance (curve 3). This necessitates the continuous monitoring of the condition of structures on site. That is, the main requirement of a monitoring and testing strategy is to measure the **'state of health'** of the building or the structure **'on completion'**, which can then be checked regularly during its **'life'** by further routine collection of data.

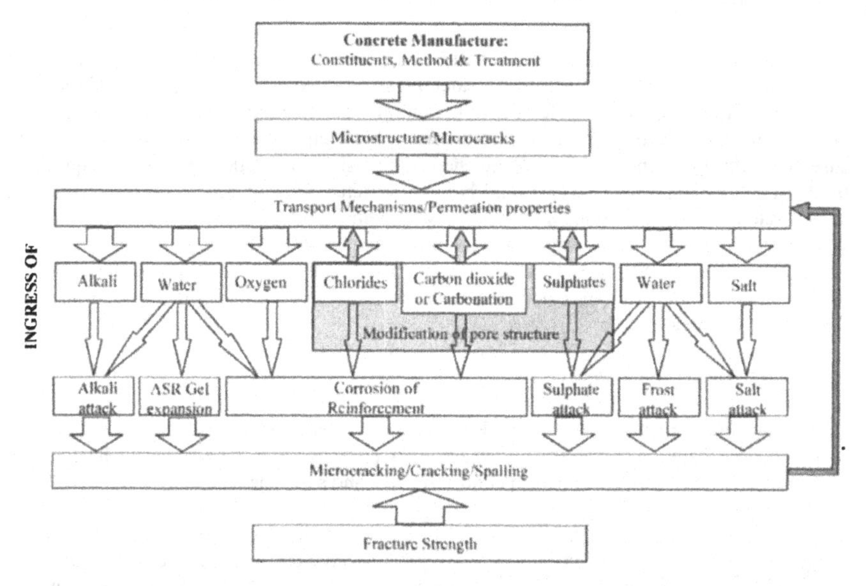

Fig. 3 Dependence of durability of concrete on microstructure and transport mechanisms

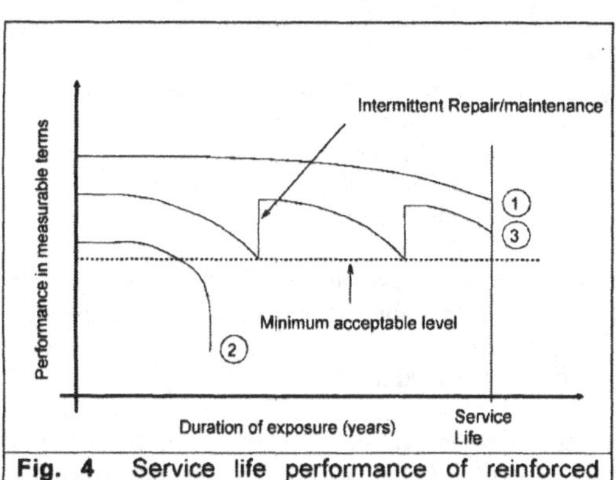

Fig. 4 Service life performance of reinforced concrete structures

The above discussion highlights that in any study of the deterioration of concrete structures, more than one type of property of concrete will be required to identify the nature and the extent of deterioration due to various causes. This is because more than one property of concrete influence many types of deterioration. Therefore, the **objectives of monitoring and testing** of concrete structures are:

⇒ to investigate the current condition of the structure

⇒ to diagnose the causes of defects or deterioration

⇒ to select an appropriate solution to the problem

In this paper, an overview of various methods which can be used to satisfy the above objectives are described. Details are given under permeation tests and sensor systems for use and application to assessing the durability of concrete structures.

PERMEATION MEASUREMENTS

The movement of gases, liquids and ions through concrete, generally called permeation, occur due to various combinations of air or water pressure differentials, humidity differentials and concentration or temperature differences of solutions [1]. Depending on the driving force of the process and the nature of the transported matter, different transport processes for deleterious substances through concrete are distinguished as diffusion, absorption and permeation. Whereas the Permit Ion Migration Test [2] could be used to determine the ionic diffusion characteristics of concrete on site, the Autoclam Permeability System [3] could be used to measure the gas and water permeability and water absorption (sorptivity) of concrete. In the following section, a brief description of these instruments is first given, followed by an illustration of the relationships between permeation properties obtained with the Autoclam Permeability System and various durability measures.

Autoclam Permeability System

The Autoclam Permeability System (Fig. 5) can be used to measure the air permeability, sorptivity and water permeability of concrete. It consists of two parts, a mechanical part to perform the test and electronic part to control the test and store the data.

Fig. 5 The Autoclam Permeability System

In order to carry out an Autoclam permeation test, an area of 50 mm diameter is isolated on the test surface with a metal ring. The ring can be either bonded to the test surface with an adhesive or clamped with a rubber ring to provide an airtight seal.

The air permeability test is carried out by increasing the air pressure on the test surface to 0.5bar and noting the decay of pressure with time. The decay of the pressure is monitored every minute for 15 minutes or until the pressure has diminished to zero. A plot of natural logarithm of pressure against time is linear and the slope of the linear regression fit of data between 5th and 15th minute for tests lasting for 15 minutes is reported as an air permeability index, with units of Ln(pressure)/minute. When the pressure becomes zero before the test duration of 15 minutes, the data from the start is used to determine the slope.

Either the water absorption (sorptivity) test or the water permeability test can be carried out at the location of the air permeability test at least 1 hour after the latter. If both the air permeability test and any of the water flow test are to be performed at the same location, it is important to do the air permeability test first because the test area will get wet with the water flow test. Also, it is to be noted that the water permeability test cannot be carried out at the same test location after the completion of the sorptivity test.

The procedure used for both the water flow tests with the Autoclam is similar; the main difference is in the test pressure for the two. In the case of the sorptivity test, water penetrating into concrete at a pressure of 0.02bar is measured whereas the pressure for the water permeability test is 0.5bar. In the former case, water is considered to be absorbed by the concrete, whereas the pressure also causes the flow in the latter case. Both the tests last for 15 minutes and the cumulative water penetration into concrete is plotted against square root of time, which is more or less linear. The slope of this graph is reported as the sorptivity index or water permeability index, as the case may be, with units of $m^3/\sqrt{minute}$. If the part of the graph between the 5th and the 15th minute is utilised, start-up errors are avoided.

Influence of In Situ Air Permeability Index on Durability Parameters

The depth of carbonation and salt scaling are related to the Autoclam permeability index in Figs. 6 and 7 respectively. For the range of mixes investigated (i.e. 27 different ordinary Portland cement concrete mixes), there is a very high degree of correlation between the permeability indices and the durability parameters. These results indicate that an early prediction of the susceptibility to carbonation, salt scaling or freeze-thaw deterioration can be obtained by carrying out an *in situ* permeability test on the concrete, when the concrete is reasonably dry.

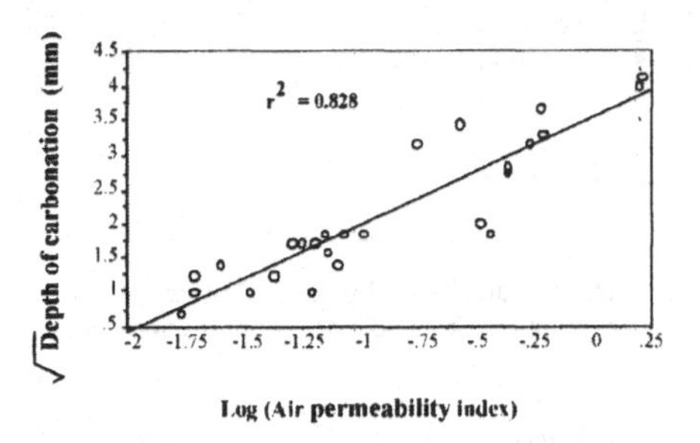

Fig. 6 Correlation between Autoclam air permeability and depth of carbonation

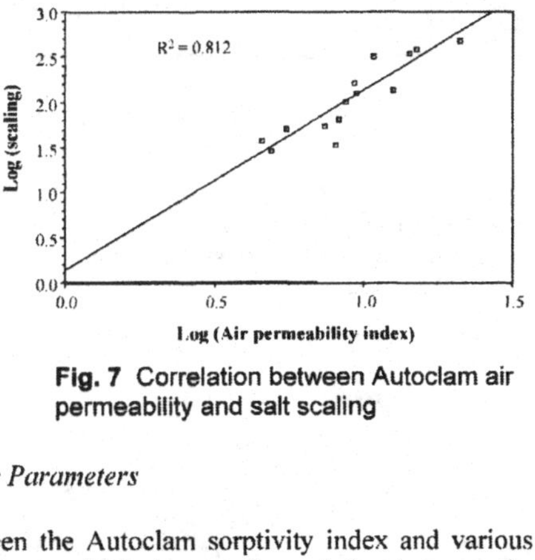

Fig. 7 Correlation between Autoclam air permeability and salt scaling

Influence of In Situ Sorptivity Index on Durability Parameters

Figures 8 and 9 present the relationships between the Autoclam sorptivity index and various durability parameters obtained. In Fig. 8, there is very high degree of correlation between the sorptivity index and the chloride penetration. This is considered to be because the chloride exposure regime for the data in Fig. 8 consisted of a wetting and drying cycle, for which the chloride penetration is depended on the absorption characteristic of the concrete. From a practical point of view, the results in Fig. 9 are very encouraging because here a very good dependence on the corrosion initiation time on the sorptivity was obtained. The effect of the depth of cover on this relationship can also be seen in this figure.

Permit ion migration test

The apparatus for carrying out the *in situ* chloride migration test is presented in Fig. 10. It consists of two cylinders, of different diameters, concentrically placed on the concrete surface, and then sealing them to prevent any flow between them along the surface of the concrete. The inner cell contains 0.55 molar sodium chloride solution and the outer cell contains distilled water. The inner cell accommodates a circular stainless steel mesh anode and the outer cell accommodates an annular mild steel perforated plate cathode. In order to carry out the test, once the apparatus is fixed on to the surface of the specimen and filled with the solutions, a potential difference of 60v dc is applied between the anode and the cathode. This forces chloride ions to travel from the anode to the cathode through the concrete in the near surface zone. After about 6 to 10 hours, depending on the quality of the concrete, a steady migration of chlorides into the outer cell is achieved, the rate of which is reported as an *in situ* chloride migration index, in mole/cm^3.s. Using the geometry of the cell and the flow path of chloride ions through the concrete, and by applying the Nernst-Planck Equation, it is, however, possible to convert this index into the chloride migration coefficient, in m^2/s [4].

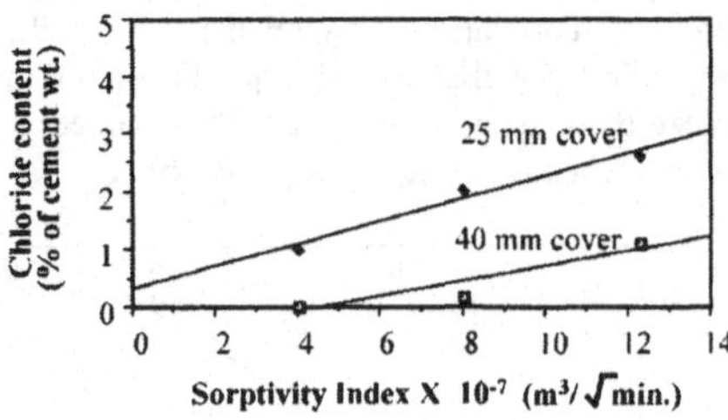

Fig. 8 Correlation between Autoclam sorptivity and chloride penetration after 10 weeks of exposure

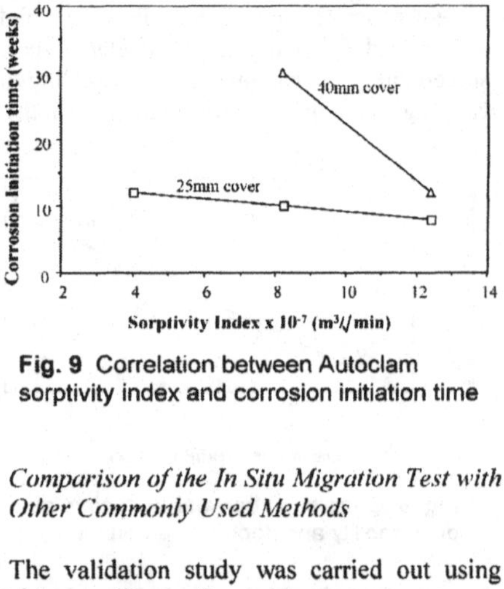

Fig. 9 Correlation between Autoclam sorptivity index and corrosion initiation time

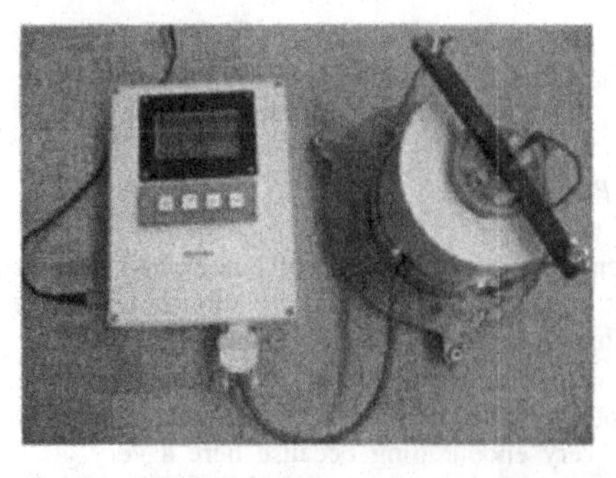

Fig. 10 PERMIT Ion Migration Test Apparatus

Comparison of the In Situ Migration Test with Other Commonly Used Methods

The validation study was carried out using eight ordinary Portland cement concrete mixes. The *in situ* migration index was compared to the migration coefficient from the one dimensional chloride migration test and the effective diffusion coefficient from the normal diffusion test. It was found that the *in situ* migration index correlated well (R^2=0.95) with the migration coefficient determined from the one dimensional chloride migration test (Fig. 11), with no zero intercept. A high degree of correlation (R^2=0.95) occurred between the in-situ migration index and the effective diffusion coefficient (Fig. 12). However, the regression line did not pass through the origin. The different tests showed that the w-c ratio is the main factor influencing the test results. The a-c ratio had little effect on both the diffusion and the migration coefficients. Therefore, it can be concluded that the Permit Ion Migration Test could be used to determine the diffusion characteristics of concrete on site. Multiple tests can be carried out simultaneously and it is also possible to use the Permit along with the Autoclam Permeability System.

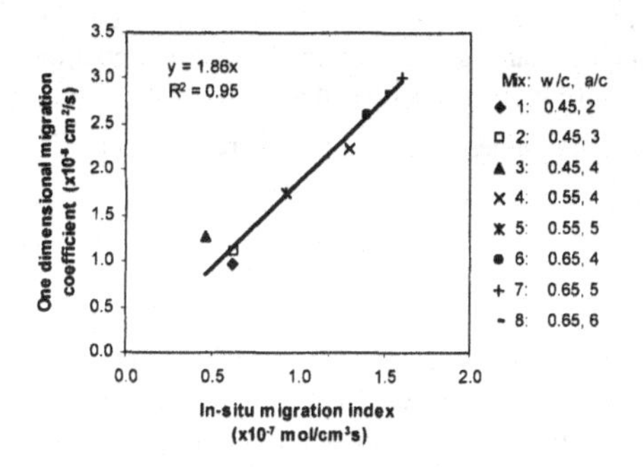

Fig. 11 Correlation between Permit in-situ ion migration index and one-dimensional migration coefficient

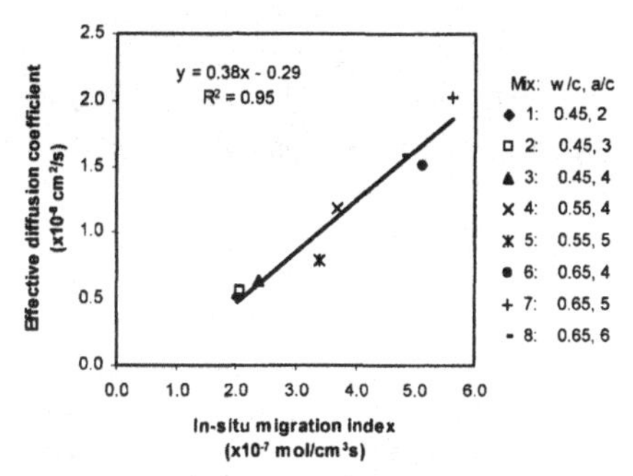

Fig. 12 Correlation between Permit in-situ migration index and effective diffusion coefficient

NOVEL SENSOR SYSTEMS

The primary function of sensor systems is to provide real-time data on the condition of the covercrete and the spatial distribution of cover-zone properties. In order to study water, ionic and moisture movement with the surface zone, two systems are noteworthy - a *multi-ring electrode moisture sensor* [5, 6] and a *covercrete electrode-array* [5, 8]. Both systems allow electrical measurements to be taken within the cover zone, although the electrode geometry and layout is significantly different between the systems. The use of the covercrete electrode array is further elaborated in this paper.

There has been progress in another type of sensor systems, which are based on optical fibre technology, to monitor pH, chloride content and moisture content [9-11]. As this is a relatively new field to most structural engineers, the background to their development and the potential for their applications to structural monitoring are elaborated in this paper.

Covercrete Electrode Array

The covercrete electrode array comprises 10 electrode-pairs mounted on a small perspex former and secured to a rebar (Fig. 13). The pairs of electrodes were mounted parallel to the suction surface enabling the electrical properties of the concrete measurements (resistance in this instance) to be obtained at 10 discrete points through the surface 50mm of concrete. Thermistors are also mounted on the former thereby enabling temperature profiles to be obtained.

McCarter et al [7] have reported covercrete electrode array data obtained from three different exposure regimes in Scotland, UK. Figure 4 shows concrete blocks of size 300x300x200mm exposed at a marine environment. The conductance (inverse of the resistance) measured across the pairs of electrodes have been presented in the following four ways:

i) variation of as-measured conductivity (in Siemens/m, S/m) as a function of time, t, for each electrode position on the sensor;

ii) variation of conductivity measurements obtained in i) which have been standardised to the reference temperature (20°C) thereby allowing changes in conductivity due to temperature to be minimised. This is particularly important for site measurements;

iii) variation of *normalised* conductivity using the values in ii) above where the normalised conductivity, N_C, is defined as,

$$N_C = \frac{\sigma_t}{\sigma_o} \qquad \text{[Eq. 1]}$$

σ_t is the standardised conductivity measured at a particular electrode position on the sensor at time, t, and σ_0 is the conductivity measured at that respective electrode position taken at a datum point in time. N_C values thus allow relative changes in conductivity to be studied. Data can also be normalised by the conductivity obtained at a depth of 50mm, i.e.

$$N^*_C = \frac{\sigma_t}{\sigma_{t,50}} \qquad \text{[Eq. 2]}$$

where σ_t is as defined above, and σ_{t50} and is the value of conductivity across the electrode pair positioned at 50mm at time, t. The 50mm electrode level was chosen as the response at this depth was found to remain relatively unaffected by drying and wetting action at the surface.

iv) variation in as-measured conductivity (or any of formalisms above) as a function of depth through the cover, thereby allowing electrical profiles through the covercrete to be studied.

The above formalisms could be applied for the resistances obtained as well. That is, instead of the normalised conductivity, it is possible to determine the normalised resistivity, as shown in Fig. 15. In this figure, the normalised resistivity is presented for electrodes at five different depths from the exposure face of a block tested in a laboratory study. As can be seen from this figure, the resistance ratio decreased with time for all the electrode pairs, but the most affected electrodes were those nearer to the exposure face.

Fig. 13 Covercrete electrode array

Fig. 14 Laboratory exposure

Figure 16 shows the conductivity values of ggbs concrete blocks exposed to the marine environment. In this figure, the as measured conductivity values, after correcting for the temperature variations, are presented. In this case, the conductivity decreased with time, i.e. the resistivity increased with time, unlike the data in Fig. 15. This indicates that not only the ggbs concrete provided resistance to the chloride ingress in the marine environment, but also the continued pozzolanic reaction had resulted in an improvement in its electrical resistance.

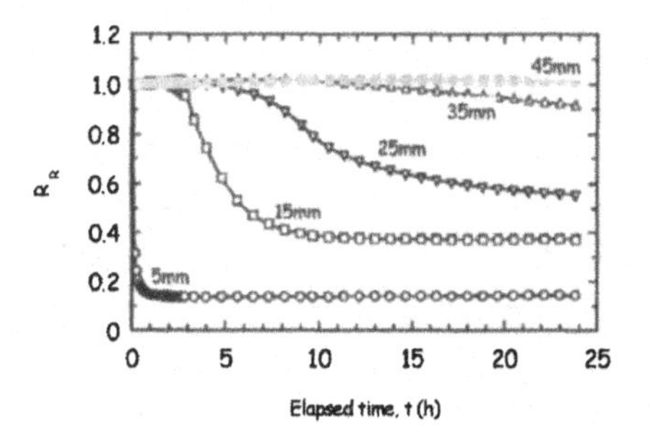

Fig. 15 Resistivity ratio for electrode pairs at different depths from the exposure face of a concrete block (OPC Concrete)

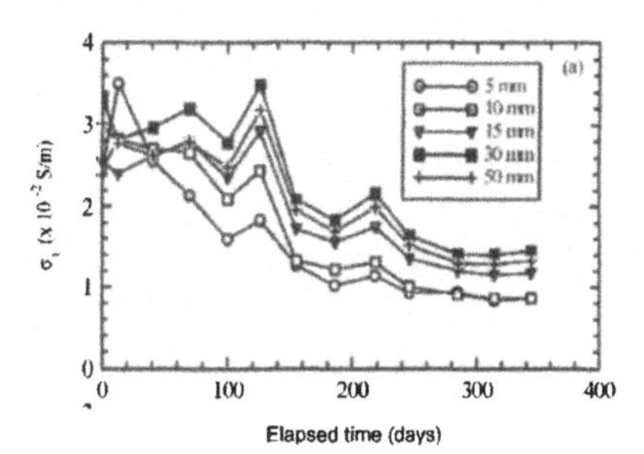

Fig. 16 Change of conductivity of ggbs concrete in the marine environment

Fibre Optic Sensors for Monitoring Moisture and chemical Changes in Structures

Optical fibre technology has developed significantly in recent years and the use of grating, especially FBG-based systems has been revolutionary for structural monitoring [10-12]. These devices can readily be retrofitted to large structures and multiplexed to simplify the input/output connections. A number of reports of bridges and related structures being instrumented with fibre optic based sensors using FBGs or, less frequently, in-line Fabry-Perot filters has been forthcoming from authors from across the world [15]. Systems typically use a network of FBGs arranged in a series of channels, continuing a range of read-out techniques to enable the large number of measurement points needed for a major structure to be achieved.

Perhaps the most exciting and scientifically stimulating of the recent developments has been the rise in interdisciplinary optical fibre sensor activities with a strong applications focus. This has been clear from recent major meetings in the field, e.g. the Optical Fiber Sensor Conference Series [16], an optical fibre sensor workshop in Spain in 2004 [17] and SPIE events such as Opto-Ireland [18]. Following this trend, the authors embarked on a research programme to develop chemical and moisture sensor systems for use in structural durability monitoring applications.

Fibre Optic Humidity Sensors

In association with a major European manufacturer, fibre optic humidity sensors have been developed and evaluated. These are based on carefully tailored coatings (typical thickness 33μm) on FBGs fabricated and then tested to respond to humidity changes and thus to monitor water ingress. The approach is to develop a sensor system to monitor the porosity of concrete structures through the change in the humidity reaching the probe, mounted centrally in a standardized concrete sample. The device has a sensitivity of 4.5pm/%RH and operates at 1535nm. The key to success has been the work done to identify, optimise and validate the response of the coatings used where they must be compatible with the conditions involved in the structure. This is shown schematically in Fig. 17.

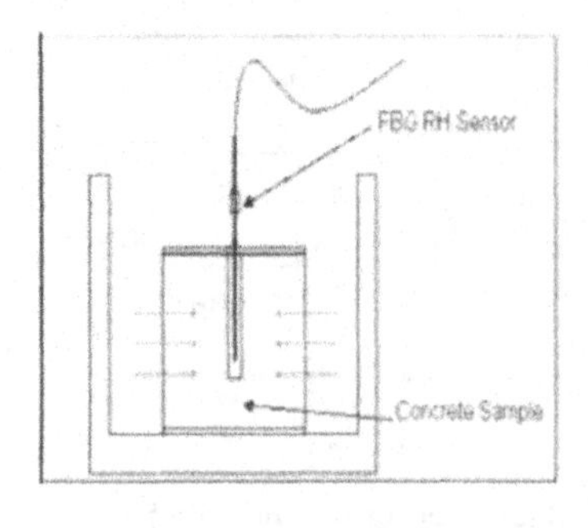

Fig. 17a Cylindrical sample with a fibre optic humidity sensor inserted centrally to monitor the ingress of water, with time, from a water bath in which the sample is placed

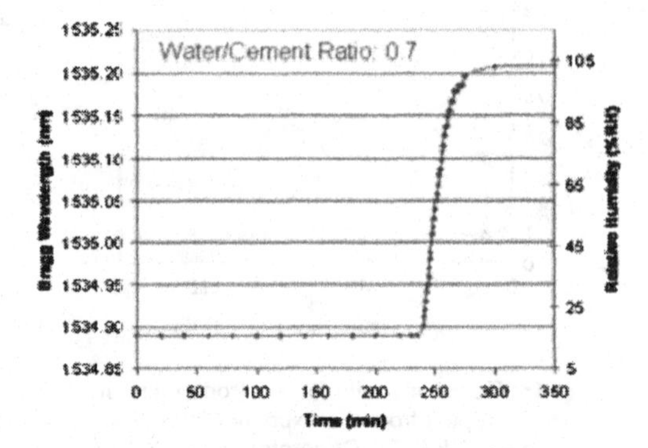

Fig. 17b Results of porosity monitoring in the cylindrical concrete sample

Fibre Optic pH Sensor System

A further important aim of recent research by the authors has been to develop sensor systems to investigate the deterioration of concrete through internal monitoring of the pH of the concrete itself. The fibre optic probes were designed to be cast into concrete samples and thus to be able to preserve their integrity in the hostile environment of the concrete. Two probe configurations were used, one using a sol-gel based probe tip into which an indicator dye has been introduced (Fig. 18a) and the second using a disc containing an indicator operating over a narrower range of pH (Fig. 18b). Both were connected to a portable spectrometer system (Fig. 19), which is used to monitor the changes in optical absorption of the probe tip. A white light source to interrogate the active elements is used as the systems operate in the visible part of the spectrum. The performance of these sensors for *in situ* monitoring of pH was tested in two different ways. In the first method, the sensors were embedded in cement mortar blocks at the time of manufacturing and they were subjected to accelerated carbonation, by which a pH profile was established in the blocks from the test surface. In the second method, after the blocks were subjected to the accelerated carbonation, the sensors were inserted in water-filled holes drilled in the blocks from the exposure surface. In order to compare the performance of the sensors, the blocks were split and sprayed with the phenolphthalein indicator solution and the depth up to the pink colouration was measured and reported as the depth of carbonation.

Figure 20 shows calibration curves for the two types of probes. It can be seen that there exists a linear relationship between absorbance and the concentration (or the pH of the solution). The sol-gel probe had a wider range than the disc probe. However, the response time varied. The disc type had a response time of 20 minutes and the sol-gel probe's response time varied depending on the thickness of the probe matrix. For a sol-gel probe with thickness less than 1.5mm, the response time was less than 5 minutes and for a thicker matrix (approximately 3mm thickness), it was about half an hour. The lifetime of the sol-gel probe is greater than three years (the duration of the current study) and it is possible to increase the lifetime by using thicker sol-gel substrates. However, the lifetime of the disc type probe by exposure to a solution of pH 12 is less than 1 month. Therefore, the disc type probe is not suitable for the long-term monitoring applications.

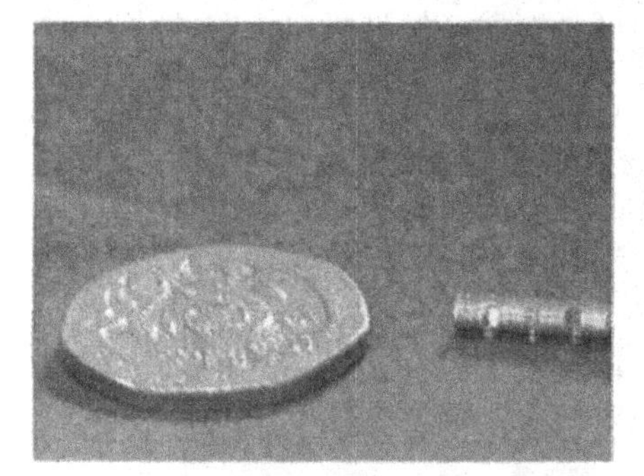

Fig. 18a FOS pH sensor using sol-gel with indicator (Cresol Red)

Fig. 18b FOS pH sensor using disc with indicator (Thymol Blue)

Fig. 19 Spectrophotometer, broadband light source, and sol-gel probe

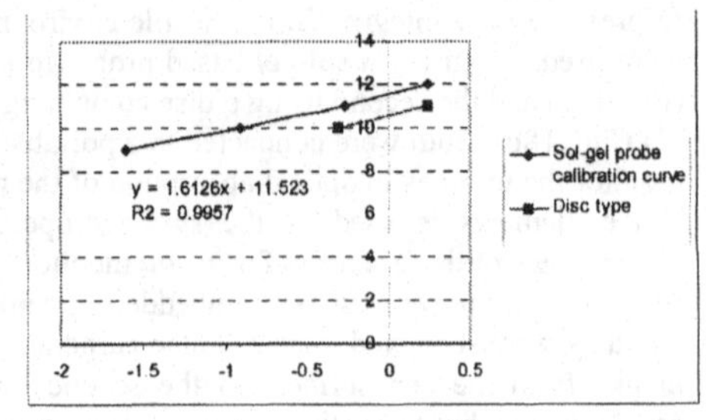

Fig. 20 Calibration curves for both sol-gel and disc type probes

As mentioned above, two sets of experiments were carried out to study the performance of the two types of sensor probes. In the first experiment, 100mm mortar cubes were subjected to an accelerated carbonation regime and the pH was measured using the demountable fibre optic sensor probes, which is presented in Fig. 21. In order to carry out measurements with the demountable type probes, 6mm diameter holes were drilled up to two depths, 5mm and 10mm, from the test surface. The depth of hole in the cubes 1 to 3 and 7 was 5mm and that in cubes 4 to 6 and 8 was 10mm. The holes were filled with water (Fig. 21) and allowed to equilibrate prior to carrying out the measurements. Cubes 1 to 6 were tested using the disc type probes and cubes 7 and 8 were tested using the sol-gel type probes. The sol-gel probe used was made from the thicker matrix. The pH values obtained using the two types of probes are presented in Tables 1 and 2 respectively. Figure 22 shows that at a time of approximately 30 minutes, the thicker type sol-gel probe was able to give a steady value of the pH of the concrete. The response time for the disc type probe was less than 10 minutes.

Fig. 21 Demountable type FOS probe

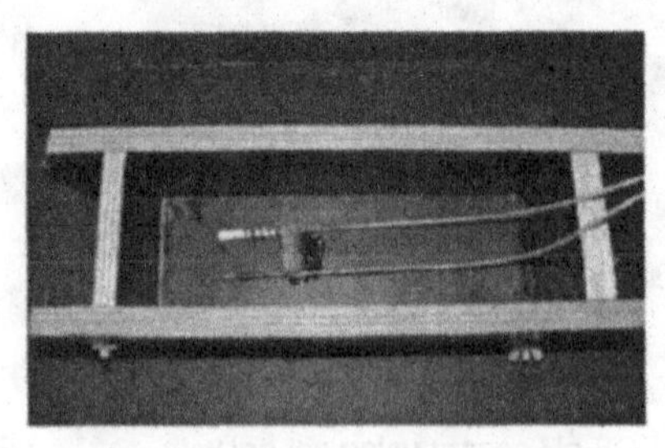

Fig. 22 The disc type (covered in white sheathing) and the sol-gel type (with metal sheathing and wrapped with a geopolymer overlay) used as embedded probes

Table 1. pH measured using the disc type FOS probes in carbonated mortar cubes

Cube No.	Cube 1	Cube 2	Cube 3	Cube 4	Cube 5	Cube 6
Depth of measurement	5mm	5mm	5mm	10mm	10mm	10mm
Disc probe	<9.0	11.8	11.5	>12.0	>12.0	12.3
pH indicator paper (for comparison)	8.0	12.0	11.0	13.0	13.0	12.5

Table 2. pH measured using the sol-gel type FOS probes in carbonated mortar cubes

Cube No.	Cube 7	Cube 8
Depth of measurement	5mm	10mm
Sol-gel probe	10.5	12.2
pH indicator paper (for comparison)	10.0	12.0

In the second experiment, embedded sensors were attached to the mould, as shown in Fig. 23. In this figure both the disc type and the sol-gel type probes can be seen. They were placed at a height of 10mm from the bottom of the mould so that they would be at a depth of 10mm from the test surface after the mould was removed. Figure 24 shows a block of cement paste embedded with the FOS. The measurement of pH commenced immediately after the mould was filled with cement mortar. From the measurement it became obvious that the disc type probe did not survive within the mortar. Fig. 25 shows the pH measured using a thin sol-gel probe during the initial 2 minutes after pouring the mortar in the mould. The sol-gel probe, however, could not measure the pH after 15 minutes of casting, presumably because the pH exceeded the upper limit of the sol-gel (i.e. 13).

The results indicated that:

(i) fibre optic pH probes can be manufactured using commercially available disc type and sol-gel sensor substrates. Both the disc type and the sol-gel type probes were able to measure the pH of the buffer solutions reliably and accurately. The disc type probe had a narrower pH range than the sol-gel type probe.

(ii) sol-gel thickness can be adjusted during its manufacturing in order to control the response time, which for the thinner matrix is about 5 minutes and for the thicker matrix is about 30 minutes. The disc type probe has a response time faster than 10 minute, but it can only be used for demountable applications.

(iii) the pH values calculated from the absorbance were close to those obtained using the pH indicator papers for carbonated cement mortar cubes. Both the disc type and the sol-gel type were able to measure the difference in pH due to the carbonation profile in the cement mortar cubes.

(iv) when embedding FOS probes in concrete, only the sol-gel probes were found to be suitable. With a thinner sensing matrix for the sol-gel probe, it was possible to obtain a faster response time, but for long-term concrete monitoring, this is not necessary.

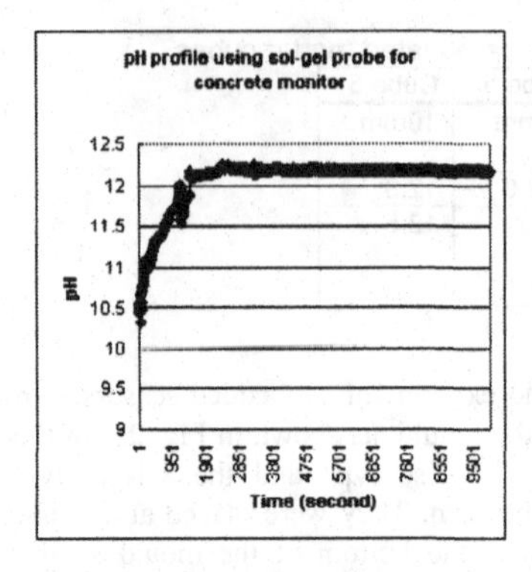

Fig. 23 pH change with time measured with the sol-gel probe in demountable application

Fig. 24. Mortar block with embedded sensors

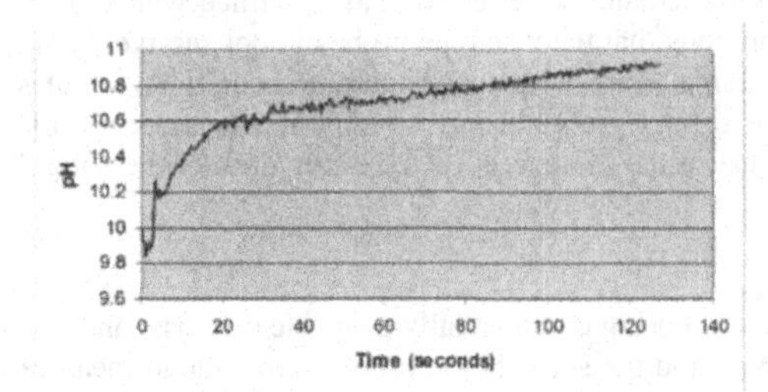

Fig. 25 Measured pH using the thin sol-gel probe during the setting of mortar samples up to 2 minutes

Fibre Optic Chloride Sensor System

The work on chemical sensors has been expanded to the study of another major mechanism for the deterioration of concrete structures, through monitoring the ingress of the chloride ions. A further sol-gel based system is under development using colour changes in silver salts as the indicator mechanism. Preliminary research, the results of which are shown in Fig. 26, has shown promising indicators in the use of a silver chromate based sensor of similar design to the sensor probe shown in Fig. 18a. The approach is to use the difference in the absorption of the species at different wavelengths and work is continuing to expand the approach to develop more stable, reversible probes for both retrofitting and fitting during construction.

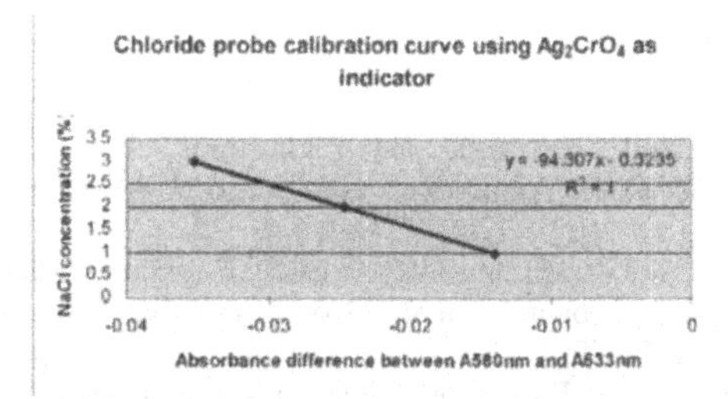

Fig. 26 Fibre optic chloride sensor based on silver chromate

CONCLUDING REMARKS

In this paper, it has been demonstrated that Autoclam permeability system, Permit Ion Migration test, the fibre optic sensor probes and the covercrete electrode array could be used to determine various physical characteristics of concrete which are directly related to its durability. There are certain advantages and disadvantages with each of these techniques, but a combination of testing and monitoring methods would ensure that properties of concrete which influence its durability are obtained right from the completion stage of a structure. Figure 27 illustrates the importance of continuous monitoring and testing for assessing the durability of reinforced concrete structures. If the properties are acceptable, they can be recorded for future use. However, as indicated in this figure, if they exceed certain threshold values, then appropriate strategies for the repair can be decided. The values can also be used to determine when a structure can be considered not fit for its intended purpose.

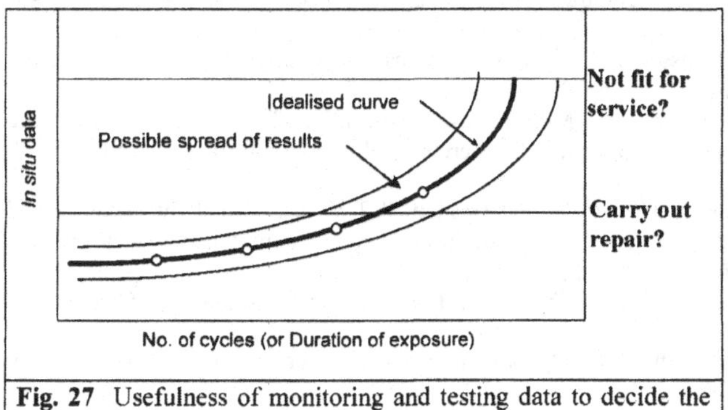

Fig. 27 Usefulness of monitoring and testing data to decide the time to repair or reconstruction

Both the electrode array and the FOS probes could be used to monitor the changes in material properties during the initiation period of deterioration. These could be supported by in-situ testing using both the Autoclam permeability system and Permit ion migration test. In the active or the propagation stage, it is highly unlikely that monitoring devices may be of any significant benefit, but in-situ test techniques could be used effectively, as illustrated by the strong correlations between permeation measurements and durability assessments. However, it is to be remembered that a combination of different types of non-destructive test techniques may be needed to assess the degree of deterioration in the propagation stage.

ACKNOWLEDGEMENTS

The authors acknowledge the support of the Engineering and Physical Sciences Research Council (EPSRC) for much of this work, together with the support of several industrial collaborators. The support of other colleagues at both City University and Queen's University is greatly appreciated.

REFERENCES

1. Basheer, P.A.M., Permeation Analysis (Chapter 16), in Handbook of analytical techniques in concrete science and technology: principles, techniques and applications, Ed: V. S. Ramachandran and J. J. Beaudoin, Barnes & Noble Publications, New Jersey, USA, 2000, October, pp. 658-736
2. Basheer, P.A.M., Andrews, R.J., Robinson, D.J., and Long, A.E., 'PERMIT' ion migration test for measuring the chloride ion transport of concrete on site, Non-destructive Testing and Evaluation International, 38, 2005, pp. 219-229.
3. Basheer, P.A.M., Long, A.E. and Montgomery, F.R., The Autoclam - a new test for permeability, Concrete, Journal of the Concrete Society, July/August 1994, pp 27-29.
4. Andrews, R.J., Design and development of an in-situ chloride migration test, PhD thesis, The Queen's University of Belfast, Northern Ireland, September 1999, 388pp.
5. Schie□l, P. and Raupach, M., Monitoring corrosion risk in concrete structures – review of 10 years experience and new developments, Proc. 5th CANMET/ACI Int. Conf. on Durability of Concrete, Barcelona, SP-192, Vol. 1, June, 2000, pp19-34.
6. Schie□l, P. and Raupach, M., Instrumentation of Structures with Sensors - Why and How?, Proc. Concrete in the Service of Mankind Conf., E&FN Spon, London, June 1996, pp1-15 (Eds. R.K. Dhir and M.R. Jones).
7. McCarter ,W.J., Emerson, M. and Ezirim, H., Properties of concrete in the cover zone: developments in monitoring techniques, Mag. Conc. Res., Vol. 47, No. 172, 1995, pp243-251.
8. McCarter, W.J. and Brousseau, R., The AC response of hardened cement paste, Cem. Conc. Res., Vol. 20, No 6, 1990, pp891-900.
9. Xie, W., Sun, T., Grattan, K.T.V., McPolin, D., Basheer, P.A.M and Long, A.E., Fibre Optic chemical sensor systems for internal concrete condition monitoring, Proc SPIE 5502, pp334-7.
10. Pal, S., Shen, Y., Mandal, J., Sun, T. and Grattan, K.T.V., Combined fluorescence and grating-based technique for wider range strain-temperature simultaneous measurement using Sb-Er doped fibre, Proc SPIE 5502, pp160-3.
11. Grattan, K.T.V. and Zhang, Z.Y., Fiber Optic Fluorescence Thermometry, Chapman & Hall, London, 1994.
12. Grattan K.T.V. and Meggitt, B.T., Optical Fiber Sensor Technology Vols 1-5, Kluwer Academic, Dordrecht, The Netherlands, 2000.
13. J M Lopez-Higuers Optical Fibre Sensing Technology published by Wiley, Chichester, England, 2002
14. Udd, E., Fiber Optic Smart Structures, Wiley-Interscience, New York, USA, 1995.
15. Doornick, J.D., Fibre Bragg Grating Sensors for Structural Health Monitoring of Civil Structures, Proc. Int. Symposium on Advances and Trends in Fiber Optics and Applications, Chongqing, China, October 2004, published by Chongqing University Press, pp293-7, 2004.
16. Optical Fiber Sensor Conferences, CD of Conference series, SPIE.
17. Lopez-Higuera, J.M. and Culshaw, B., Second European Workshop on Optical Fibre Sensors, Santander, Spain Proc SPIE 5502, 2004.
18. Opto Ireland Conference, Dublin, Ireland, April 2005, Proc SPIE 5826A 2005

INFLUENCE OF PRODUCTION METHOD AND CURING CONDITIONS ON CHLORIDE TRANSPORT, STRENGTH AND DRYING SHRINKAGE OF TERNARY MIX CONCRETE

Mateusz Radlinski and Jan Olek
Purdue University, School of Civil Engineering
550 Stadium Mall Drive
West Lafayette, IN 47907

Anthony Zander
Indiana Department of Transportation, Materials and Test Division
120 South Shortridge Road
Indianapolis, IN 46219

Tommy Nantung
Indiana Department of Transportation, Research Division
1205 Montgomery Road
West Lafayette, IN 47906

ABSTRACT

The main objective of this study was to evaluate the influence of different production methods i.e., laboratory and field mixing and placement, on the durability-related properties of concrete. Chloride transport, as determined by rapid chloride permeability (RCP) method, and drying shrinkage (free and restrained) were studied on the specimens obtained under laboratory and in-situ conditions. The latter specimens were cast during the two-phase (fall and spring) construction project of bridge deck in northern Indiana, which resulted in two different curing conditions of the field concrete. In addition, specimens were collected during the trial batch demonstration performed prior to the actual construction of the bridge deck. The same ternary mixture composition (incorporating fly ash and silica fume) was utilized in all four cases (laboratory and field trial batches and two field placements). In order to monitor the influence of the actual field exposure conditions on resistance to chloride ion penetration of bridge deck concrete, several cores were retrieved from the part of the deck placed in the fall and tested for the ability to resist chloride penetration using RCP method. Furthermore, chloride concentration profiles were determined after the first winter season. High coulomb values obtained for concrete placed in late fall were confirmed by elevated chloride levels present in the cores. The effect of curing temperature on obtaining desired level of compressive strength was minor in comparison with its impact on permeability.

INTRODUCTION

A rapidly increasing interest in utilizing high performance concrete (HPC) in applications where harsh environment is present during the entire service life of the structure has been observed for over a decade. The term "high performance" usually refers to durability characteristics, such as low permeability, high freeze-thaw resistance or low shrinkage. Bridge decks are one example of the types of structures for which these properties are crucial from the standpoint of the service life. To enhance their durability and ensure superior performance, ternary cementitious systems incorporating mineral admixtures, such as silica fume and fly ash as a replacement for cement, are often employed. Prior to construction of the actual structure, a mix design is typically developed in the laboratory. Based on the study conducted under laboratory conditions, which are somewhat idealized, it is often anticipated that a certain mixture will have similar properties when implemented in the field.

However, due to disparities in the production, placement and curing methods, the properties of concrete in the actual structure can be significantly different from those obtained from the laboratory trials. To start with, the single term "production method" encompasses several distinctive operations, such as mixing, handling, placement, compaction and finishing, each of which can greatly influence the performance of in-place concrete. The second, but not less relevant, factor influencing the laboratory and in-field concrete performance is curing, comprising of two elements: moisture and temperature. As far as moist curing is concerned, even a 7 days curing period recommended for some bridge decks[1], is substantially shorter than continuous moist curing usually used in the laboratory. Considering the curing temperature, in turn, it should be remembered that laboratory specimens are cured under constant temperature of 23 ± 2°C, a condition which is very different from that likely to be encountered in the field. Providing sufficiently high curing temperature at early age is particularly critical for concrete mixtures with high content of supplementary cementitious materials such as fly ash or slag due to their reduced reactivity at low temperatures[2].

This study was focused on assessment of impact of production method of ternary mix concrete on chloride transport properties, shrinkage and compressive strength. Furthermore, the influence of curing conditions on these properties was evaluated.

OBJECTIVES OF THE STUDY

The primary goal of this study was to compare performance of HPC of the same nominal mix design prepared under field and laboratory conditions. The performance parameters under evaluation were: permeability (measured using RCP method), the depth of chloride penetration, compressive strength development, and free and restrained shrinkage. The second objective was to assess the impact of curing temperature on these properties, as well as the effect of moisture availability during the curing period on the development of compressive strength.

Project overview

This research project, aiming at development of optimum ternary concrete mixture for bridge decks applications, is associated with the recent construction of HPC concrete bridge deck on SR-23 in South Bend, Indiana. Construction was divided into two phases, each consisting of two lanes. Phase I bridge deck construction (BDC-1) took place in early November of 2004, whereas phase II construction (BDC-2) was conducted during the early May of 2005. From the technological point of view, the only difference between the two construction phases was that the BDC-1 concrete was placed using a pump, while the conveyor belt (in some portions) or the crane and a bucket were used to place BDC-2 concrete. Apart from acceptance specimens collected as per INDOT's QC/QA procedures, additional specimens were cast in order to assess the various properties of concrete. Furthermore, to verify the quality of in-situ concrete, two sets of cores were retrieved from BDC-1 phase. In addition, prior to the actual construction, the key properties of the proposed ternary concrete mixture were established during the field trial batch demonstration (FTB) carried out by the ready-mix concrete producer that was selected to supply concrete for the actual construction. The concrete mixture of the same composition was also produced in a laboratory (LTB).

EXPERIMENTAL PROGRAM

Mix composition and materials used

The criteria related to the concrete design (CMD) set out on for the SR-23 project were as follows[3]:

- The paste volume of total cementitious material and water shall not exceed 28% of concrete volume design value.
- The cement content in the ternary system shall be at least 231 kg/m^3 of concrete.
- Class F or C fly ash shall be used as part of the total cementitious materials content in the ternary binder system. Fly ash shall constitute 20 to 30 percent by mass of total cementitious materials content in the mix design.
- Silica fume shall constitute 5-7 % of the total cementitious materials content in the mix design.
- The water-cementitious materials ratio shall be no less than 0.38 and not exceed a maximum of 0.42.
- The CMD target air content shall be set at either 6.5% or 7.0%, with a 100% pay factor range of 5.9-8.9 %.
- The slump shall be within a range of 100 mm to 190 mm.
- The target compressive strength at 28-days shall be a minimum of 33.8 MPa.

The selected mix proportions are presented in Table I. The binder system contained ordinary portland cement (ASTM C150 Type I) combined with 20 % of class C fly ash and 5 % of densified silica fume, by weight of total cementitious materials (332 kg/m^3). The physical properties and chemical composition of the cementitious materials are provided in Table II. The nominal water to cementitious materials ratio (w/cm) was 0.40, while the paste content was 24% by volume.

Table I. Concrete mixture composition

Material	Cement	Fly Ash (Class C)	Silica Fume	Fine Aggregate (SSD)	Coarse Aggregate (SSD)	Water
Content (kg/m^3)	249	66	17	778	1062	133

All four series of mixtures described in this paper were produced using the same source of materials. The only exception was the high-range water reducer (HRWR) used in LTB mixture, which was provided by the different manufacturer than the HRWR used in all other mixtures. The coarse aggregate used was crushed limestone with specific gravity (SSD) of 2.71 and absorption of 1.14 %. The fine aggregate used was natural siliceous sand having specific gravity of 2.63 and absorption of 1.24 %. Both fine and coarse aggregate were stockpiled separately for this project, and essentially the same material was used in both construction phases as well as in field and laboratory trial batches. For the LTB mixture the coarse aggregate was sieved and recombined to ensure the same gradation as that of the stockpiled material.

Table II. Physical properties and chemical composition of cementitious materials

Description of test	Cement	Fly Ash	Silica Fume
Physical tests:			
Specific gravity	3.15	2.67	2.20
Fineness – passing 325 mesh	99	9.6	–
– Blaine's surface area (cm^2/g)	3620	–	–
Compressive strength tested on mortar cubes (psi) – 1 day	2310	–	–
– 3 day	3890	–	–
– 7 day	4870	–	–
– 28 day	6060	–	–
Chemical analysis:			
Silicon Dioxide SiO_2 (%)	20.60	36.16	93.07
Aluminum oxide Al_2O_3 (%)	4.70	20.32	0.62
Ferric oxide Fe_2O_3 (%)	2.60	7.58	0.41
Calcium oxide CaO (%)	64.90	23.94	0.66
Magnesium oxide MgO (%)	2.60	5.47	1.16
Sulfur oxide SO_3 (%)	2.50	1.91	<0.01
Loss on ignition (%)	1.31	0.40	2.71
Total alkali as sodium oxide Na_2O (%)	0.53	1.35	0.67
Insoluble residue (%)	0.34	–	–
Bogue potential compound composition:			
Tricalcium silicate C_3S (%)	65	–	–
Dicalcium silicate C_2S (%)	10	–	–
Tricalcium aluminate C_3A (%)	8	–	–
Tetracalcium aluminoferrite C_4AF (%)	8	–	–

Tests conducted and testing methods used

Standard QC/QA tests, such as slump, air content, and unit weight were performed on fresh concrete. To evaluate the resistance of concrete to chloride ion penetration, the rapid chloride permeability (RCP) test (AASHTO T277) was used in the study. Despite its frequent criticism[4], this test method is believed to provide a useful comparative measure of concrete resistance to chloride penetration[5, 6, 7], which was one of the goals of this study. The reported results are based on the average obtained from four 52 mm thick disks, two sampled from the top part of the 102×203 mm cylinder (after discarding the top 6.5 mm) and two sampled from the bottom part of a cylinder (114-165 mm from the top). The RCP test results reported for the cores retrieved from the bridge deck were obtained from the 51×95 mm disks after removal of the top 12 mm.

The free shrinkage test was performed on three 76×76×286 mm prismatic specimens following the procedure of AASHTO T160. The restrained shrinkage test, frequently called the ring test, was conducted in accordance with AASHTO PP34, except that the non-standard ring geometry was used (Figure 1). The test method involved casting of concrete ring, with cross-sectional dimensions of 76×76 mm, in a cardboard mold placed around 10 mm thick steel ring of 305 mm external diameter, which was instrumented with four strain gages placed symmetrically at the mid-height. 24 hours after casting the outer cardboard mold was removed and the top and bottom surfaces of the ring were sealed with aluminum tape. As a result, only the outer circumference of the ring was undergoing drying. Two replicate specimens were cast in the case of FTB and BDC-1 mixtures, while only one ring was fabricated for LTB. A third ring was also cast to monitor temperature developing in BDC-1 concrete, as shown in Figure 1.b). For comparison purposes, the paper also presents shrinkage data collected directly from the bridge deck poured during the first construction phase. These data were obtained from two vibrating wire strain gages located in the deck; one gage was placed on the reinforcing bar located in the mid-span of the deck at equal distance from the girders, and the second gage was embedded in the surrounding concrete. Compressive strength tests were carried out on three cylinders at each testing age. 102×203 mm cylinders were utilized in the case of BDC-1, BDC-2 and LTB, while for FTB 152×305 mm specimens were used. Finally, the chloride concentration profiles were determined for the 94 mm diameter cores retrieved from the bridge deck constructed during the phase I construction following the ASTM C1556 procedure.

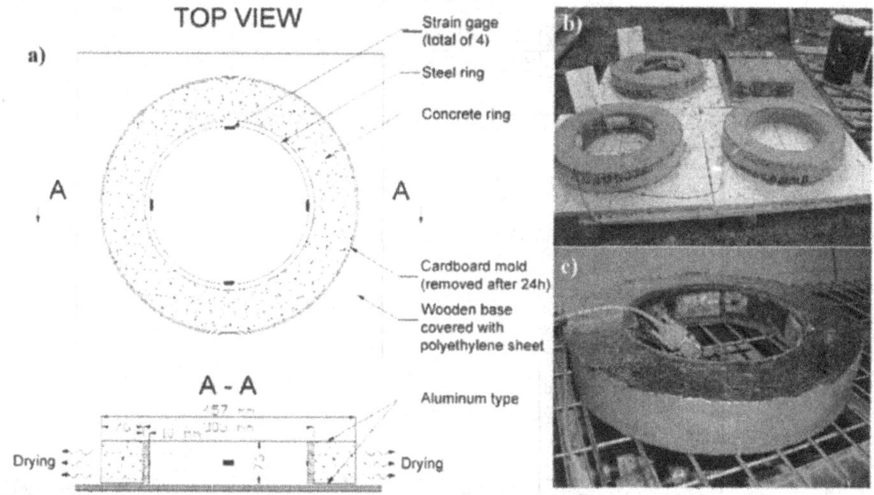

Figure 1.a) Restrained shrinkage test setup; b) BDC-1 ring test specimens after casting; c) LTB ring test specimen

Curing conditions

The effect of curing conditions on the examined properties of HPC was of major interest in this work. Focus was made on the impact of the early age temperature on permeability and compressive strength for BDC-1 and BDC-2, as well as on the influence of moisture availability on the strength development for the FTB. The detailed description of curing regimes for respective tests for each case of the study is provided in Table III.

The BDC-1 compressive strength specimens were exposed to the simulated field conditions until the testing time. The conditions included 7 days of moist curing under wet burlap covered with plastic sheet. RCP specimens, however, were field exposed until the 35th day, at which time they were split into three groups. The first group of specimens remained exposed to the in-situ environment, the second group was moved to the lab (23°C, 50% RH) and the last group was moved to the moist room (23°C, 100% RH). The ring specimens, in turn, remained exposed to the field conditions throughout the entire testing period.

In the case of BDC-2, both compressive strength and RCP specimens were divided into two groups from the very beginning. The first group underwent a standard moist curing and the other one was subjected to the field exposure after 7 days of curing under wet burlap covered with polyethylene sheet.

FTB and LTB specimens for RCP test were cured in a standard way, i.e. moist curing at 23°C until testing. While LTB specimens for compressive strength were cured in a standard way as well, the FTB specimens were subjected to three different curing regimes, namely moist curing, air drying (23°C, 50% RH), and intermittent moist curing (saturation on day 3, 5 and 9 for six hours, otherwise air drying). The ring specimens for restrained shrinkage test were cured with under burlap for 7 days, followed by storage at about 65% RH and 50% RH, for FTB and LTB respectively. Free shrinkage specimens were in all cases moist cured for 7 days, which corresponded to the curing period of the actual structure, followed by exposure to 50% RH.

Table III. Curing conditions for each test and study case

Test \ Source of specimens	BDC-1	Cores from BDC-1 (actual structure)	BDC-2	FTB	LTB
RCP	1) 7 days under wet burlap covered by plastic sheet followed by field exposure*	7 days of wet burlap + plastic sheets followed by exposure to ambient conditions	1) 7 days of wet burlap and plastic sheet followed by field exposure	Lab moist cured	Lab moist cured
	2) As in 1), from 35th day-on air cured @ 23.0°C and 50% RH		2) Lab moist cured		
	3) As in 1), from 35th day-on air cured @ 23.0°C and 100% RH				
Compressive strength	7 days of wet burlap and plastic sheet followed by field exposure*		1) 7 days of wet burlap and plastic sheet followed by field exposure	1) Lab moist cured	Lab moist cured
			2) Lab moist cured	2) Lab cured @ 23.0°C and 50% RH (air drying)	
				3) Intermittent lab moist cureing (6 hours on day 2, 5 and 9)	
Free shrinkage	Lab moist cured, from 7th day-on @ 23.0°C and 50% RH	Cores obtained @ 149 and 303 days	Lab moist cured, from 7th day-on @ 23.0°C and 50% RH	Lab moist cured, from 7th day-on @ 23.0°C and 50% RH	Lab moist cured, from 7th day-on @ 23.0°C and 50% RH
Restrained shrinkage (ring test)	4 days under wet burlap covered by plastic sheet followed by field exposure*		-	Lab cured with wet burlap and plastic sheet, from 7th day-on @ 23.0°C and 65% RH	Lab cured with wet burlap and plastic sheet, from 7th day-on @ 23.0°C and 50% RH

*Note: The term field exposure refers to storage of specimens at the job site for 24 hours followed by outdoor storage in the courtyard of concrete laboratory of Purdue University, IN (about 100 miles south-east from the job site). Due to very similar air temperature cycles recorded at both locations (Figure 2), the exposure conditions in the courtyard were assumed to be comparable to that experienced at the construction site.

RESULTS AND DICUSSION

Temperature history data

Presented in the Figure 2 is the in-place temperature history for the BDC-1 concrete as well as the ambient temperature at the job site and at the storage (after transferring specimens from the job site). It can be noticed that the ambient temperature for specimens transferred from the job site to the courtyard of concrete laboratory at Purdue University was very similar to the actual ambient temperature at the job site. This supports a claim of previously mentioned similarities of field exposure (see note under Table III). The data also indicated that throughout the first 10 days after casting the concrete temperature was significantly higher than the ambient temperature due to heat liberated during early stages of hydration. Coincidently, after about 10 days, the temperature in the deck dropped below 0°C and from that time-on remained slightly below the ambient temperature.

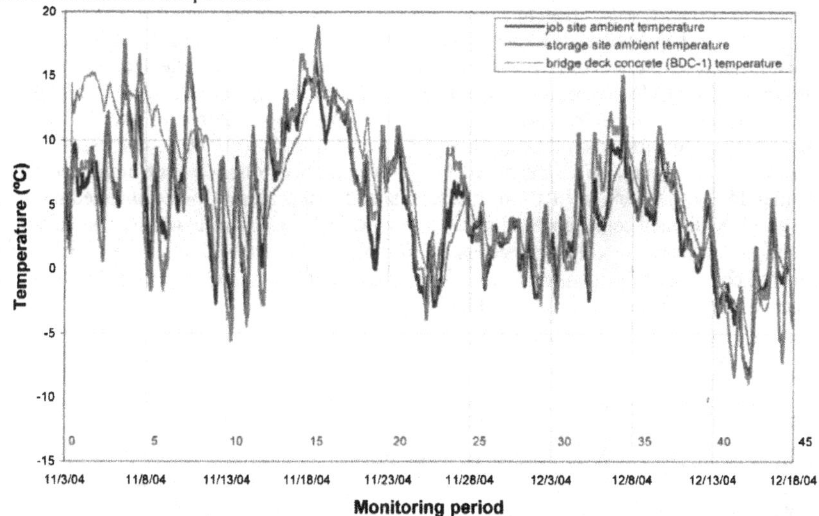

Figure 2. Temperature history for BDC-1 concrete[8] throughout the period of initial 45 days

Fresh concrete properties

Figure 3 contains the test results obtained from the fresh concrete. It can be seen, that in all cases the measured slump fell well within the required range of 100÷190 mm. Some problems were encountered with providing air content within the recommended range of 5.9÷8.9 % during the first phase of construction, and in one of the sublots sampled the air content was determined to be 3.9%. This was a result of pumping, which dramatically altered the air content throughout the entire phase I construction. The measured unit weight was in all cases within 1% of the theoretical value (calculated based on the measured air content), indicating accurate batching of mixing water. This is evident from the opposite trend in the air content and unit weight results (Figure 3).

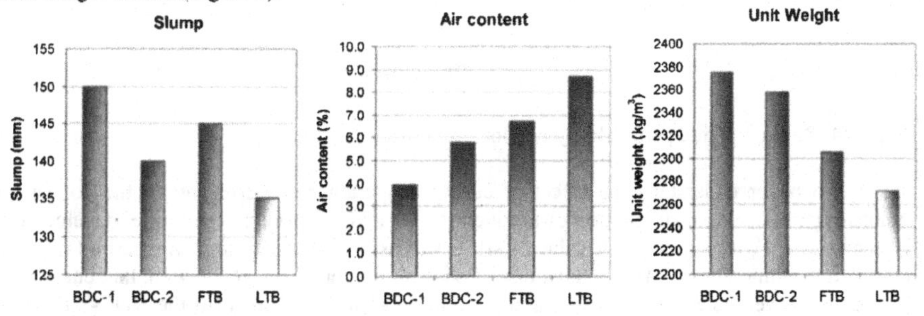

Figure 3. Fresh concrete properties

RCP test results

The graphs presented in Figure 4 clearly illustrate large variation in the RCP results obtained for different cases studied. The best results were obtained for the BDC-2 concrete cured under field conditions. Moist cured laboratory trial batch concrete achieved similar coulomb values at a given time, while moist cured BDC-2 concrete yielded slightly higher permeability. The FTB concrete, also moist cured until the testing time, achieved relatively high result at 28 days (about 2500 coulombs), but significant drop occurred by the 56th day, and commonly used for bridge decks requirement of maximum 1000÷1500 Coulombs at 56 days[3] was satisfied. However, BDC-1 exposed to field conditions exhibited very high permeability even after 150 days. Very low RCP results (average of 520 coulombs) were obtained only after 303 days for additional cores retrieved from the bridge deck. Also, worth noticing is the fact that transferring specimens to either laboratory or to moist room at the 35 days resulted in a very substantial coulomb reduction already at 56th day (Δc_{lab} and Δc_{moist} respectively). This reduction can most likely be attributed to increased hydration rate.

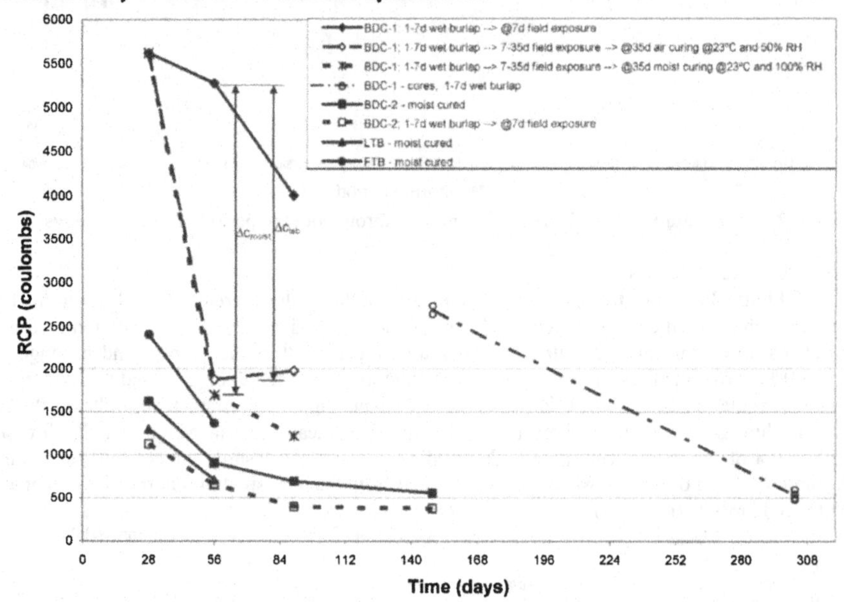

Figure 4. Rapid chloride permeability results

To better understand the effect of curing temperature on permeability, the concept of maturity index, being a summation of product of time and temperature over time, developed by Nurse and Saul[9], was utilized. A value of -10°C was assumed as a datum temperature (T_0). The cumulative temperature plotted against time is shown in Figure 5. It can be seen, that cumulative temperature over time for LTB has parallel and almost identical trend to the BDC-2 concrete exposed to the field conditions. This explains well very similar RCP test results obtained for these two groups of specimens. Moreover, it can be noticed that due to low curing temperature it took BDC-1 concrete over 200 days to obtain cumulative temperature corresponding to that of LTB or BDC-2 concrete attained at 90 days. Likewise, substantial increase in cumulative temperature can be observed for BDC-1 specimens which were transferred to the most room at 35th day, reflecting the already mentioned rapid drop in RCP test results (Figure 4). In general, it appears that maturity, primarily used for prediction of compressive strength, may be applicable to a wide spectrum of concrete properties[10, 11].

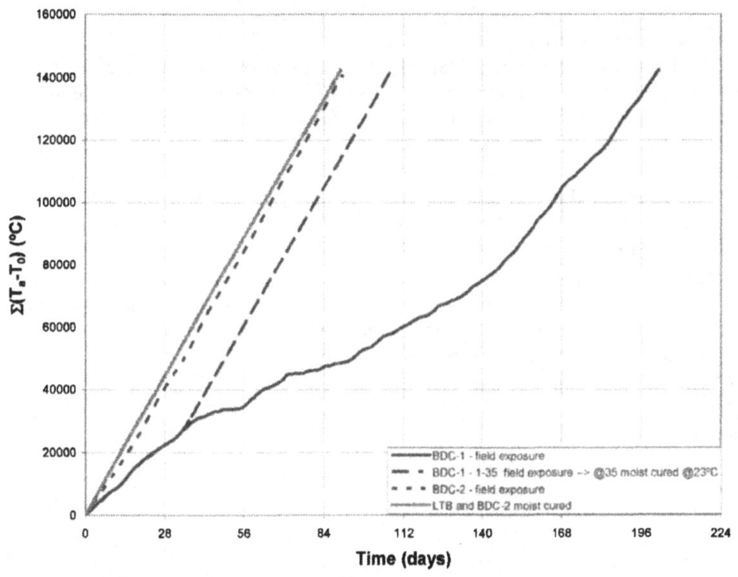

Fig.5. Cumulative temperature over time

The question arises why BDC-2 moist cured specimens exhibited higher RCP results than the ones exposes to the field conditions, given that he temperature was almost identical (Figure 5). This apparent discrepancy can be explained by the fact that the standard AASHTO T277 procedure requiring 19 hours of saturation (including one hour of vacuum saturation) does not provide the same saturation if the specimens were continuously moist cured as opposed to being air dried. This is illustrated in Table IV showing comparison of 150 days RCP results accompanied by the mass gain and saturation time of BDC-2 specimens cured in a standard way and exposed to the field conditions for two preparation procedures. The first procedure involved the standard preparation, whereas in the second specimens for both curing conditions were soaked until no change in mass over time was recorded. It can be seen that when completely saturated, the apparent difference in permeability for the two curing regimes diminishes (490 coulombs for moist cured specimens vs. 477 coulombs for field exposed specimens), which confirms the expected results based on the almost identical cumulative temperature encountered.

Following this view, it should be realized that the BDC-1 specimens which were also air drying under field exposure, would have in fact even higher results if they had been brought to the same saturation level as moist cured specimens.

Table IV. Comparison of 150-day RCP results for different preparation procedures

	Standard procedure		Full saturation	
	moist cured	field exposed	moist cured	field exposed
RCP (coulombs)	554	376	490	477
Mass gained during saturation (g)	1.7	8.2	3.0	15.4
Saturation time after completion of vacuum sat.	18 hours		27 days	

Chloride concentration profiles

The total of two cores were obtained for determination of chloride concentration profiles after the first winter from the bridge deck constructed during phase 1. The first core was retrieved from the driving lane, the second core was extracted from the gutter located along the temporary concrete barrier. Water-soluble chloride content is shown in Figure 6. The relatively high level of chloride near the surface, that occurred after one winter season only, is attributed to relatively high permeability of the ternary concrete mixture which was exposed to deicing salt only 21 days after construction. The average ambient temperature during that period was about 7°C (Figure 2). For comparative purposes, also shown in Figure 6 are the data for other bridge decks located in the area of Fort Wayne, Indiana[12]. OPC concrete of w/c = 0.45 with no mineral admixtures was used for these structures. Chloride profiles were obtained from those bridge decks at 3, 4, 5, 7 and 9 years. Noteworthy is the fact that despite high permeability, ternary mix concrete demonstrated very good chloride binding capacity.

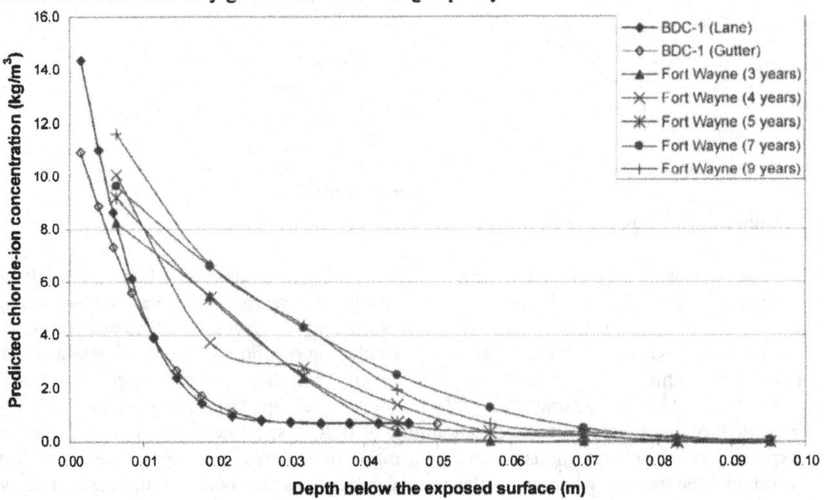

Figure 6. Chloride concentration profiles; Fort Wayne data from ref.[12]

For comparison purposes, diffusion coefficient and surface concentration were determined for the analyzed chloride profiles (Table V) using Fick's second law and the method of least squares. The effects of other phenomena affecting the chloride transport, such as, wicking action, sorption, and permeation have been omitted in the analysis.

Table V. Diffusion coefficient and surface concentration

	Bridge deck location						
	BDC-1 - Lane	BDC-1 - Gutter	Fort Wayne Area[12]				
Exposure time (years)	0.34		3	4	5	7	9
C_s (kg/m^3)	16.50	12.33	10.57	12.33	11.53	11.21	13.78
D_a (10^{-12} m^2/s)	3.776	5.274	3.607	2.096	2.089	2.941	1.579

Compressive strength comparison

Despite having the highest air content of all concretes tested (see Figure 3) , and therefore being in the position to have its strength most adversely affected[13], concrete produced under laboratory conditions developed the highest strength. This might be explained by more precise batching and more thorough mixing taking place in the laboratory. Furthermore, affecting the strength development is initial temperature within first 24 hours[14]. It is also important to notice that, unlike resistance to penetration of chloride ions, compressive strength was not as sensitive to low temperature curing, and therefore attaining desired levels of strength at 28 days did not cause difficulties (Figure 7). Similar findings were obtained by Khan and Ayers[15], as well as Bentur and Jaegermann[16]. Both of these studies, concluded that the adverse effect of inadequate curing conditions at early age is more pronounced in the case of concrete permeability than in the case of strength.

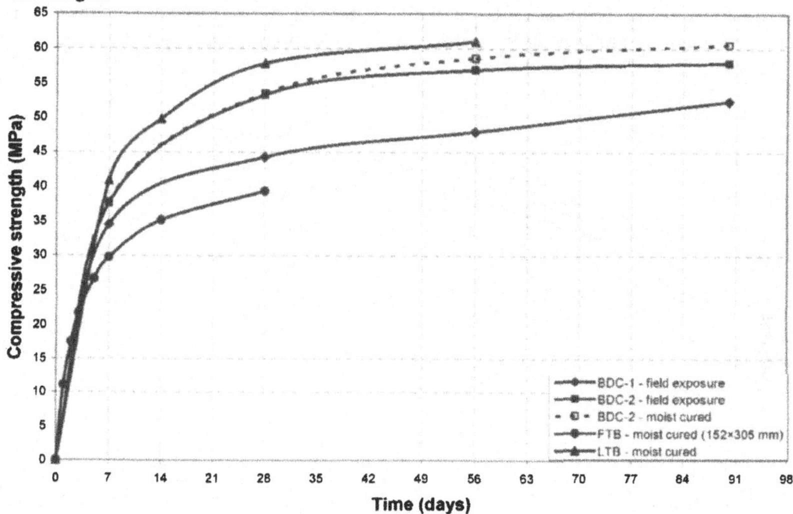

Figure 7. Compressive strength development

It should be noted that with an exception of FTB concrete compressive strength was tested on102×203 mm cylinders. Since, for FTB mixture, 152×306 mm cylinders were used, a correction of about 5÷10% of strength at a given age should be considered in order to directly compare the test results[14]. Such correction will results in the upward shift of the entire strength curve. Considering relatively large variation in air content of different mixes, the values of compressive strength seem to be quite comparable.

Influence of moisture availability on compressive strength development

The FTB specimens cast for compressive strength were divided into three groups, each subjected to different curing regime. The intermittent moist curing, being between the two extremes, such as moist curing and air drying, represents a situation in which concrete is subjected to non-continuous moisture supply. It can be seen that up to three days the curing regime did not affect strength at all (Figure 8). However, starting from the fifth day a progressive relative loss of strength with respect to moist cured specimens can be observed especially in the case of specimens subjected to air drying. For this treatment, the time-strength curve flattens down very rapidly, and no strength gain was encountered after 14 days. These results indicate that in order for hydration to progress, certain supply of water is necessary. Furthermore, even relatively limited amount of moisture provided during early age, can help to develop strength comparable with that of continuously moist cured concrete.

Similar conclusion can be drawn on the basis of previously mentioned Figure 7 clearly illustrating the influence of moisture availability on compressive strength by comparison of moist cured and field exposed BDC-2 specimens cured at almost the same temperature (Figure 5). It can be noticed that specimens which were moist cured under wet burlap for seven days only, developed similar compressive strength to those which were continuously moist cured. Negligible difference which occurred from 56 days-on became slightly more pronounced at 90 days. This proves that, with respect to compressive strength, seven days of moist curing is sufficient and due to formation of relatively impermeable and discontinued pore structure, further curing does not result in significant increase in compressive strength.

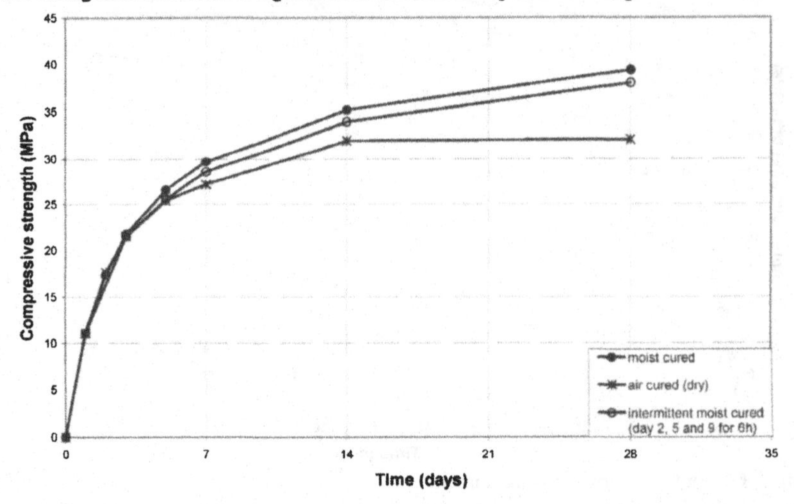

Figure 8. Compressive strength development for FTB as a function of moisture availability

Shrinkage

Similar shrinkage characteristics were obtained for BDC-1, FTB and LTB concretes (Figure 9). Much lower strains, however were developed in concrete produced during the second phase of construction (BDC-2). This unusual departure from data collected for the other cases can be explained by the fact, that as mentioned previously, concrete was placed in May on a relatively warm and sunny day. Even though shrinkage prisms were covered with wet burlap and plastic sheet, water might have evaporated quickly, which resulted in premature drying of the specimens. Upon transferring them from the job site to the laboratory, approximately 24 hours after casting, there were no visible signs of moisture on the surface. Using the standard procedure, the specimens were soaked in saturated lime water about 45 minutes before taking the measurements, which might have not resulted in very effective re-saturation. As a result, the initial readings were not taken in the state of sufficient moisture content, and accordingly the readings taken after subsequent six days of moist curing yielded an abnormal expansion (about 145 με). It should be noticed that if the shrinkage curve for BDC-2 was shifted down by 120 microstrains, leaving much more reasonable 25 microstrains for the expansion effect, the strains at 119 days would be 430 microstrains, or very close to results obtained for other cases. This shows that lack of moisture provided during the initial 24 hours period may potentially lead to misleading results and false conclusions.

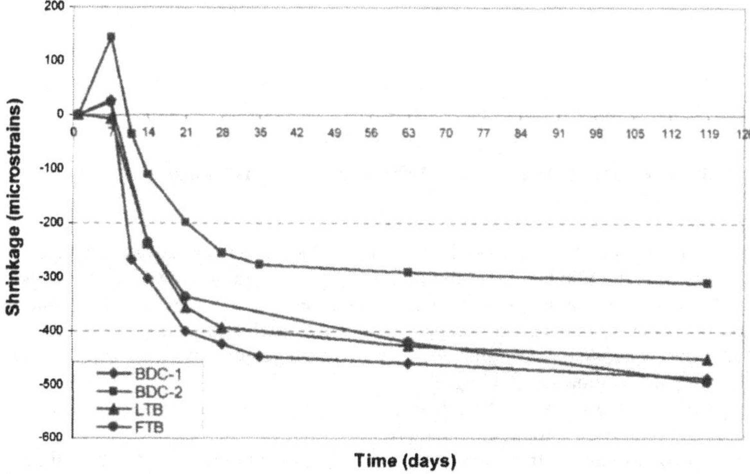

Figure 9. Free shrinkage over time

Restrained shrinkage test carried out for the BDC-1, FTB and LTB concrete showed several interesting features (Figure 10). Firstly, due to combination of insufficient restraint level provided by the non-standard steel ring and relatively low shrinkage level, none of the rings cracked within the testing period. Secondly, the maximum strain levels attained reflects the relative humidities at which the specimens were stored during the test, i.e., 50%, about 65%, and 77% on average, for LTB, FTB and BDC-1 respectively. Furthermore, considerable fluctuation in the strains can be observed for the BDC-1 concrete which occurred due to ambient temperature and relative humidity changes. Finally, it can be seen that the ring test results for BDC-1 are relatively close to the actual strains in the bridge deck measured with two strain gages, one located on the rebar and another embedded in the concrete. Therefore, it can be concluded that the restrained shrinkage test results represent strain level which is very comparable with that which might occur in the actual structure, as long as the test specimen is subjected to similar relative humidity.

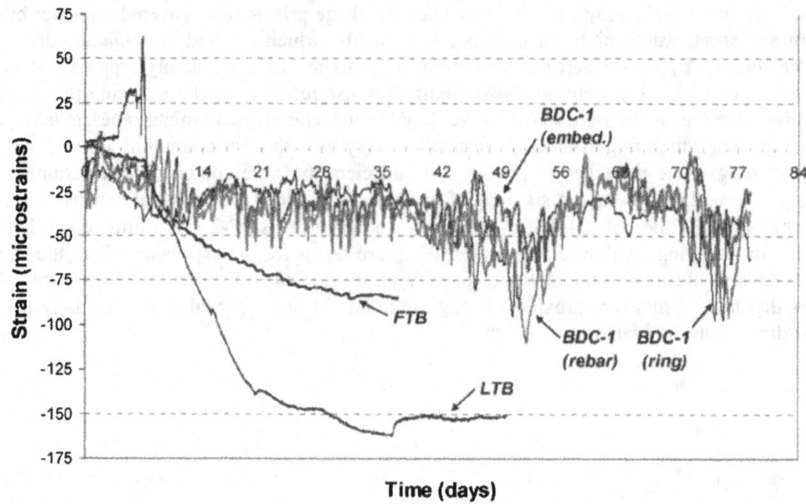

Figure 10. Restrained shrinkage over time, BDC-1 (rebar and embedded) data from ref.[8]

CONCLUSIONS

The following conclusions have been drawn on the basis of conducted study:

- In the ternary cementitious system investigated, the early age curing temperature had strong influence on resistance to chloride ion penetration. Over time, low-temperature cured concrete developed permeability comparable to that of high-temperature cured ternary mixes.
- Ternary mix concrete placed and cured at low temperature was capable of developing sufficient 28-day compressive strength.
- No significant difference in shrinkage characteristics was observed between the mixes prepared using different production methods.
- Mixture produced and cured under laboratory conditions exhibited only slightly better performance than mixes produced in the field.

REFERENCES

[1]T. C. Holland, "Practical considerations for using silica fume in field concrete," *Transportation Research Record*, **1204**, pp. 1-7 (1988).

[2]A. Krishnan, J. Mehta, J. Olek, and W. J. Weiss, "Technical Issues Related to the Use of Fly Ash and Slag During Late Fall (Low Temperature) Construction Season, FHWA/IN/JTRP Draft Final Report (2004).

[3]"Quality Control/Quality Assurance, QC/QA, Superstructure Concrete, Modified," Indiana Department of Transportation Provisional Specification for SR-23 Bridge (2004).

[4]C. Shi, "Another look at the rapid chloride permeability test (ASTM C1202 or AASHTO T277). " FHWA Report, (2003).

[5]R. Bleszynski, R. D. Hooton, D. A. Thomas, and C. A. Rogers, "Durability of Ternary Blend Concrete with Silica Fume and Blast-Furnace Slag,": Laboratory and Outdoor Exposure Site Studies, *ACI Materials Journal/September-October 2002*, 499-508.

[6]K. D. Stanish, R. D. Hooton and D. A. Thomas, "A rapid migration test for evaluation of the chloride penetration resistance of high performance concrete," *Proceedings of the PCI/FHWA International Symposium on High Performance Concrete*, Orlando Florida, 358-367 (2000).

[7]P.C. Äitcin, "The durability characteristics of high performance concrete: a review," *Cement & Concrete Composites,* **25**, Elsevier, 409-420 (2003).

[8]T. S. Aldridge, "Structural Behavior of High-Performance Concrete Bridge Decks", M.S. Thesis, Purdue University, 182 p. (2004).

[9]A. G. A. Saul, "Principles underlying the steam curing of concrete at atmosphere pressure," *Magazine of Concrete Research*, March 1951, 127-140.

[10]J. Zhang, D. Cusson, L. Mitchell, T. Hoogeveen, and J. Margeson, "The Maturity Approach for Predicting Different Properties of High-Performance Concrete," *Proceedings of the Seventh International Symposium on the utilization of High-Strength/High-Performance Concrete*, ACI, 135-154 (2005).

[11]R. J. Kehl, C. A. Constantino, and R. Carrasquillo, "Match-Cure and Maturity: Taking Concrete Strength Testing to a Higher Level," *Project Summary Report 1714-S*, Center for Transportation Research, The University if Texas at Austin (1998).

[12]J. Olek, F. Rajabipour, A. Lu., X. Feng, A. Zander, and T. Nantung, "Influence of Mixture Composition on the Predicted Service Life and Chloride Transport Properties of Concrete," *Proceedings of Advances in Cement and Concrete Conference*, D. A. Lange, K. L. Scrivener, J. Marchand, Eds., Copper Mountain, Colorado, August, 313-328 (2003).

[13]M. Hale, S. F. Freyne, B. W. Russel, "Entraining Air in High-Performance Concrete and Its Effect on Compressive Strength," *Seventh International Symposium on the utilization of High-Strength/High-Performance Concrete*, ACI, 173-188 (2005).

[14]M. Plante, G. Cameron, A. Tagnit-Hamou, "Influence of Curing Conditions on Concrete Specimens at Construction Site," *ACI Materials Journal/March-April 2000*, 120-126.

[15] S. M. Khan, M. E. Ayers, "Curing requirements of silica fume and fly ash mortars," *Cement & Concrete Research*, **23**, Elsevier, 1480-1490 (1993).

[16]A. Bentur, C. Jaegermann, "Effect of curing and composition on the properties of the outer skin of concrete," *ASCE Journal of Materials in Civil Engineering*, **3**, No. 4, 252-262 (1991).

ACKNOWLEDGEMENTS

The authors gratefully acknowledge Gary Mithoefer, Youlanda Belew and Kevin Brower from INDOT for their contribution to this study. Help with instrumentation and data collection for the restrained shrinkage test received from Timothy Aldridge and Prof. Robert Frosch from Purdue University is also greatly appreciated. Finally, thanks are also extended to Adam Rudy, Zhifu Yang and Karol Kowalski for their help with field specimens preparation.

Appendix A - Workshop Program

Session A. Case histories & Current initiatives

Session Chair: Jacques Lukasik, Senior Vice President, Scientific Affairs LAFARGE, Paris France

State DOT Best Practices for Minimizing Concrete Bridge Deck Cracking and Enhancing Deck Performance
Lou Triandafilou, FHWA Resource Center, Baltimore, MD

Concrete Pavement Durability
Leif Wathne, American Concrete Pavement Association, Washington DC.

Properties of Aging Concrete in Dams
Tim Dolen, US Bureau of Reclamation, Denver, Co.

The Mechanisms Of Corrosion: The Effects Of Fabrication, Exposure, And Interaction With Other Materials
Ronald Latanision, Piotr Moncarz, Exponent – Failure Analysis Associates.

Session B. Fundamentals of concrete as a porous medium

Session Chair: S.P. Shah, Walter P. Murphy Professor of Civil Engineering Director, Center for Advanced Cement-Based Materials, Northwestern University,Evanston, IL

A Multi-Technique Investigation of Nanometer Structure of Calcium Silicate Hydrate
Julie Gevrenov, Jeffrey Thomas, Hamlin Jennings, Georgios Constantinides and Franz-Josef Ulm
Civil and Environmental Engineering, Materials Science and Engineering Northwestern University, Evanston IL, and Civil and Environmental Engineering, Massachusetts Institute of Technology, Cambridge MA

Ten Observations from Experiments to Quantify Water Movement and Porosity Percolation in Hydrating Cement Pastes
Dale Bentz
Building and Fire Research Laboratory, NIST, Gaithersburg, MD

Measuring Permeability and Bulk Moduli Using the Dynamic Pressurization Technique
D.A. Lange, G.W. Scherer, Z.C. Grasley, J.J. Valenza
Univ. of Illinois at Urbana-Champaign, Urbana, IL, and Civil & Env,. Princeton Univ., Princeton, NJ.

Session C. Test Methods, Parameters, and Properties

Session Chair: Jan Olek, Professor of Civil Engineering, Purdue University West Lafayette, IN

Role of In-situ Testing and Monitoring in Assessing the Durability of Reinforced Concrete
P.A.Muhammed Basheer, Adrian E. Long, Daniel McPolin, Lulu Basheer, Kenneth V. Grattan, Tong Sun, W. John McCarter, Queen's University Belfast, School of Civil Engineering, Belfast, Northern Ireland, UK, City University, School of Engineering and Mathematical Sciences, London, UK, Heriot-Watt University, School of Built Environment, Edinburgh, UK,.

Validation of Service-Life Models Using Field Data,
Michael Thomas, University of New Brunswick, CANADA.

On Tensile Strength of Hardened Portland Cement - Key-Factor for Improving Durability
Jan Van Mier, Swiss Federal Institute of Technology, Zürich, Switzerland

Session D. Durability Mechanisms: Parameters & Opportunities

Session Chair: Neal Berke, Research and Development Fellow, W.R. Grace, Construction Products Division, Cambridge, MA

Alkali Silica Reactivity of Silica Fume Agglomerates
Maria C. Garci Juenger, University of Texas, Austin, TX

Mitigation of Alkali Silica Reaction: A Mechanical Approach
Claudia Ostertag, University of California, Berkeley, CA

Mitigation of Alkali-Silica Reaction (ASR) of Highly Reactive Aggregates with Class C fly ash
James K. Hicks, Mineral Resource Technologies, Inc, The Woodlands,TX
Chemistry and Expansion in Alkali-Silica Reaction
Leslie Struble, University of Illinois Urbana-Champaign, Urbana, IL

Session E. Modeling Transport Properties, Interaction of Transport and Cracking

Session Chair: David Lange, Professor, Associate Department Head and Director of the Center of Excellence for Airport Technology, University of Illinois, Urbana-Champaign

Putting Scientific Rigor into Concrete Durability Tests: Thoughts and Benefits
Edward Garboczi, Inorganic Materials Group, NIST, Gaithersburg, MD

Cracking and Transport Property
S. P. Shah, Northwestern University, Evanston, IL

Chemo-physical and mechanical approach to performance assessment of structural concrete and soil foundation
Koichi Maekawa, Tetsuya Ishida, Ken-Ichiro Nakarai, University of Tokyo, Tokyo, Japan

Restraint and Cracking during Non-uniform Drying of Cement Composites
John Bolander, University of California-Davis, Davis, CA

Session F. Modeling Transport Properties, Physical and Chemical Aspects

Session Chair: Dale Bentz, Chemical Engineer, Building and Fire Research Laboratory, National Institute of Standards and Technology, Gaithersburg, MD

Solid Solution Model for Calcium Silicate Hydrate Applied to Cement Degradation using the Continuum Reactive Transport Model FLOTRAN,
J. William Carey, and Peter C. Lichtner, Los Alamos National Laboratory, Los Alamos, NM

Predicting the long-term durability of concrete exposed to chemically-aggressive environments – Influence of temperature fluctuations
J. Marchand, E. Samson, and T. Zhang, CRIB-Department of Civil Engineering, Laval University, Sainte-Foy, Canada, SIMCO Technologies Inc., Québec, Canada.

Modeling Approach to Dense Packing of Concrete Aggregates
K. Sobolev, and A. Amirjanov, R. Hermosillo*, Universidad Autónoma de Nuevo León, San Nicolas, NL, México, Near East University, Nicosia, N. Cyprus.

Session G. Innovative Test Methods, Parameters, and Properties

Session Chair: Edward Garbozi, Leader, Inorganic Materials Group, NIST

The Benefit of Quantifying Inherent Variability for Use in In situ Sensor Interpretation and Numerical Prediction for Service-Life Models
Thomas Schmit , Aleksandra Radlinska, Jason Weiss, Purdue University, West Lafayette, IN

Influence of production method and curing conditions on chloride transport and drying shrinkage of ternary mix concrete
Mateusz Radlinski, Jan Olek, Anthony Zander and Tommy Nantung
Purdue University, School of Civil Engineering, West Lafayette, Indiana Dept. of Transportation, Research Division, West Lafayette, IN, and Indiana Department of Transportation, Materials and Test Division, Indianapolis, IN

Direct Phase-Resolved Strain Measurements in Cementitious Materials
Joseph J. Biernacki, Sean E. Mikel and R. Wang, Tennessee Technological University, Cookeville, TN, Georgia Institute of Technology, Atlanta, GA

Experimental Studies on Coupling Effects between Moisture and Chloride Diffusion in Concrete
Lydia Abarr, A. Suwito, Ayman Ababneh, and Yunping Xi, Department of Civil, Environmental and Architectural Engineering, University of Colorado, Boulder, CO, Department of Civil and Environmental Engineering, Clarkson University, Potsdam, NY.

Session H. Field implementation of Durability Concerns: Mechanisms: Parameters, Problems & Opportunities

Session Chair: Jan Skalny, Consultant

Modeling of Stiffness Degradation and Expansion in Cement Based Materials Subjected to External Sulfate Attack
B. Mobasher, Arizona State University

2003/2004 National Survey Results on State DOT High Performance Concrete Implementation
Lou Triandafilou, FHWA Resource Center, Baltimore, MD

Session I. Summary Discussion: Parameters, Problems & Opportunities

Moderators:

Geoff Frohnsdorff	NIST, (Emeritus)
Della Roy	Arizona State University

A general discussion of Workshop topics, involvement of various code agencies, development of a research, and implementation agenda

Appendix B – Workshop Sponsors

Program Sponsors
- Department of Civil and Environmental Engineering, Arizona State University
- ASU Pavements and Materials Conference Committee
- Lafarge Corporation
- Grace Construction Products
- Salt River Project
- NIST

Session Sponsors
- Arizona Cement Association

Reception Sponsors
- Salt River Materials Group
- Degussa Admixtures
- Forta Corporation
- MACTEC Engineering & Consulting, Inc

Author Index

Printed in the United States
By Bookmasters